Mobile Platforms and Development Environments

Synthesis Lectures on Mobile and Pervasive Computing

Editor

Mahadev Satyanarayanan, *Carnegie Mellon University*

Mobile Platforms and Development Environments
Sumi Helal, Raja Bose, and Wendong Li
2012

Quality of Service in Wireless Networks Over Unlicensed Spectrum
Klara Nahrstedt
2011

The Landscape of Pervasive Computing Standards
Sumi Helal
2010

A Practical Guide to Testing Wireless Smartphone Applications
Julian Harty
2009

Location Systems: An Introduction to the Technology Behind Location Awareness
Anthony LaMarca and Eyal de Lara
2008

Replicated Data Management for Mobile Computing
Douglas B. Terry
2008

Application Design for Wearable Computing
Dan Siewiorek, Asim Smailagic, and Thad Starner

2008

Controlling Energy Demand in Mobile Computing Systems

Carla Schlatter Ellis

2007

RFID Explained: A Primer on Radio Frequency Identification Technologies

Roy Want

2006

Mobile Platforms and Development Environments

Sumi Helal, Raja Bose, and Wendong Li

ISBN: 978-3-031-01355-3 paperback

ISBN: 978-3-031-02483-2 ebook

DOI 10.1007/978-3-031-02483-2

A Publication in the Springer series
SYNTHESIS LECTURES ON MOBILE AND PERVASIVE COMPUTING

Lecture #9
Series Editor: Mahadev Satyanarayanan, *Carnegie Mellon University*
Series ISSN
Synthesis Lectures on Mobile and Pervasive Computing
Print 1933-9011 Electronic 1933-902X

Mobile Platforms and Development Environments

Sumi Helal
University of Florida

Raja Bose
Nokia Research Center, North America Lab

Wendong Li
Nokia, Location & Commerce

SYNTHESIS LECTURES ON MOBILE AND PERVASIVE COMPUTING #9

ABSTRACT

Mobile platform development has lately become a technological war zone with extremely dynamic and fluid movement, especially in the smart phone and tablet market space. This Synthesis lecture is a guide to the latest developments of the key mobile platforms that are shaping the mobile platform industry. The book covers the three currently dominant native platforms—iOS, Android and Windows Phone—along with the device-agnostic HTML5 mobile web platform. The lecture also covers location-based services (LBS) which can be considered as a platform in its own right. The lecture utilizes a sample application (TwitterSearch) that the authors show programmed on each of the platforms.

Audiences who may benefit from this lecture include: (1) undergraduate and graduate students taking mobile computing classes or self-learning the mobile platform programmability road map; (2) academic and industrial researchers working on mobile computing R&D projects; (3) mobile app developers for a specific platform who may be curious about other platforms; (4) system integrator consultants and firms concerned with mobilizing businesses and enterprise apps; and (5) industries including health care, logistics, mobile workforce management, mobile commerce and payment systems and mobile search and advertisement.

KEYWORDS

mobile platforms, smart phones, tablets, iOS, Android, Windows Phone, mobile web, HTML5, iPhone, history of mobile platforms, mobile application developments, location based services, LBS, mobile maps, mobile ecosystem

Contents

Preface

This lecture presents the key platforms shaping the mobile and portable devices industry. Mobile platforms have become a hotspot with extremely dynamic and fluid developments, especially in the smart phone market. The authors had to revise the contents of this lecture several times in less than a year and half in response to the frequent developments. The authors believe an online synthesis, which allows for extensive follow-on changes and amendments, is in fact the most suitable form in which this subject should be published.

The lecture presents the latest on key mobile platforms, utilizing a sample application (TwitterSearch) to demonstrate how it can be programmed on each of the platforms.

We hope the lecture is informative and useful, and hope to be able to extend it and keep it up-to-date over the next few years. If interested, please subscribe to the "Mobile Platforms" Google Group dedicated for this lecture, to send us your comments, feedback and corrections, and also to learn of any updates or Errata that we may post in the future. The group Home Page is: `http://groups.google.com/group/mobile-platforms`

Sumi Helal, Raja Bose, and Wendong Li

February 2012

CHAPTER 1

From the Newton to the iPhone

The first mobile information platform was probably a piece of stone or clay with markings on it, used for recording numeric information. As languages and the writing technology developed, human civilization progressed onward to the papyrus scroll and ink pen. With the invention of printing by early 1800's, the printed book became the preferred mobile platform for accessing standardized information in a portable format.

Pen and paper and books remained the platforms of choice until the late 20th century. Even though recording devices (photographical and audio recorders) were invented over the past 100 years, they were not mobile platforms for the masses.

1.1 RECENT HISTORY OF MOBILE PLATFORMS

The history of mobile platforms took a steep turn forward and picked up significant speed when the first Personal Digital Assistant (PDA) was introduced by Apple in 1987. The Apple Newton was born, marking the beginning of a two decades marathon and an amazing journey towards the iPhone historical marker (Figure 1.1).

While Newton was not a great success commercially, it nevertheless set the vision and succeeded in starting the PDA endeavor leading to today's mobile platforms. The Newton attempted to realize a Tablet concept by utilizing a stylus with a handwriting recognition as the main user interface. The stylus idea started a short period of "pen computing" developments and products that peaked and then faded into extinction, all within 20 years. The Tablet idea lived on, however, and returned in the form of the iPad (Figure 1.1). The Newton OS consisted of: (1) a microkernel for what was then considered a low-power ARM RISC architecture; (2) operating system services written in C++ offering PDA manager services and handwriting recognition; and (3) a NetwonScript application development environment, which was an object-oriented language.

As the then mobile platform de jour, Motorola used Newton OS to develop the next thing—a wireless Newton. The Motorola Marco and its executive version, the Envoy (shown in Figure 1.2), were introduced in 1995, utilizing the CDPD wireless data service (Cellular Digital Packet Data - a technology that offered packet data over circuit-switched networks).

Neither the Newton nor the Marco/Envoy lived long enough to evolve into several generations like the iPhone. In fact, only 4,000 units of the Envoy were sold before the product was discontinued. It was too early for the consumers and even the developers to grasp the full potential of these earliest PDA developments. Also, the relatively high cost of these early PDAs did not help. And so the Newton and the Envoy hit the "pen and paper" wall and crashed.

The Apple Newton
lauched in 1987
The Apple iPhone & iPad
launched in 2007 & 2010

Figure 1.1: Apple Newton as a PDA and a Table concept, redone two decades later as the iPhone and the iPad.

Figure 1.2: Motorola's Envoy – A Wireless Newton.

One culprit that was suspected back then as being responsible for the failure of the Newton and Envoy was the form factor of the PDA. Palm-sized devices (primarily, the Palm Pilot) were introduced just when the Newton was shutting down. These devices used the Palm OS developed by then startup Palm Inc., which was quickly acquired by US Robotics. A couple of years later, the handheld form factor came out with many devices, almost all of which used the Windows CE operating system (which marked Microsoft's entrance into the PDA and mobile platform arena). The Tablet followed, utilizing Windows CE or the Windows desktop OS. At that time (1990–2000), there were simply too many hand-helds, palms, and other form factors (e.g., the Vadem handheld/tablet combo) on the market (Figure 1.3). Each had hoped to discover the proper form factor, and of course, to gain part of the market share.

Figure 1.3: A Zoo of Handhelds, Palms and Tablets.

Parallel to the PDA endeavor, the "Communicator" and smart phone concepts evolved from within the telecommunications industry.

Introduced in 1996, The Nokia Communicator was perhaps as clueful a development as the Newton, and was the concept that bridged the gap between PDA's and digital cellular telephony. The Communicator combined PDA features with the then de facto networked applications (e.g., Telnet), phone and fax, all within one clam-shell design that folds a wide rectangular display onto a QWERTY keyboard (Figure 1.4). With the emergence of the Blackberry and other smartphones, the Communicator mission came to an end with a final model in 2007—the birth year of the iPhone.

First Nokia Communicator 9000 (1996, GeOS, 397g)

Fifth & last Nokia Communicator (2007, Symbian OS, 210g)

Figure 1.4: First and Last Nokia Communicator.

1.2 FIRST GENERATION MOBILE PLATFORMS

Smart phones started to enter the market place at a fast pace by the late 1990's. Palm OS, Symbian, Windows Mobile (the predecessor of Windows Phone) and BlackBerry OS were the most dominant platforms at that time, and now considered the first generation of mobile platforms. We briefly trace the history of Palm OS and Symbian developments.

The Palm OS offered a simple, single-tasking environment with strong support to handwriting recognition, and extensive personal information management (PIM) support. Applications can be developed in C or C++. Palm abandoned its own OS in 2005, in a sudden decision to switch to and adopt the Windows Mobile platform for Palm devices. Palm threw another surprise in 2009 when it announced a come back to inventing and using its own platforms, this time offering webOS – an interesting mobile platform designed entirely around web technology. Under webOS, all applications can be developed using HTML, Cascading Style Sheets (CSS) and JavaScript. A year later, Hewlett Packard acquired Palm and announced its plans to launch the TouchPad, a product that HP sadly killed shortly after it hit the market place.

Symbian OS evolved from Psion's EPOC, which was created in the mid 1990's. It consists of the Symbian OS core and three UI frameworks to choose from: S60, UIQ and MOAPS. MOAPS was developed and used by

NTT DoCoMo, while the S60 and UIQ were licensed to a few companies including LG, Motorola, Nokia, Samsung and Sony Ericsson. In 2008, Nokia acquired Symbian, Ltd., and immediately announced plans to form the Symbian Foundation to provide royalty-free open Symbian source mostly based on the Symbian Core and the S60 UI. Symbian, as a first-generation smartphone OS, achieved great success and dominated the market by a big margin for several years. In 2010, Gartner predicted that Symbian will continue to dominate the market in 2014. As it turned out, Android surpassed Symbian in 2011 although Symbian is still one of the top mobile OSes. The current Symbian version is known as the Symbian Belle. Its UI is completely overhauled and rebuilt to match other modern mobile platforms. The preferred development platform of Symbian Belle is Qt—an open-source cross-platform application framework acquired by Nokia from Trolltech.

1.3 J2ME AND BREW

In 1999, Sun Microsystems' Java 2 Platform Micro Edition (J2ME) was introduced as an open standard and lightweight virtual machine platform for mobile phones and other devices such as set-top boxes. J2ME was perceived as the next big development in mobile platforms, given the popularity and power of Java and the inherent promise of writing once, running everywhere. Two years later, Qualcomm introduced BREW, its Binary Runtime Environment for Wireless, which competed with J2ME in a strategically necessary step to keep the CDMA based mobile phone market competitive.

In a way, J2ME and BREW were the first serious attempts at creating standardized computing and networking "platforms" out of mobile phones. Original equipment manufacturers (OEMs) rapidly adopted these platforms, promising cellular operators additional ARPU (operator lingua and short for Average Revenue Per User) out of the potential mobile applications that could be developed on their handsets.

But returns on investments out of J2ME and BREW were hardly achieved. Consumer apps, including games, did not do very well, even though enterprise applications (company directory, simple CRM, field force automation, etc.) performed better in the U.S. market. In the end, J2ME and BREW mobile platforms did not take off despite the power of Java and BREW and the promise they created.

There are several reasons that led to this modest impact of J2ME and BREW.

- Unlike what Sun Microsystems has intended, J2ME was never a fully *write once run everywhere* enabler. The evolution of the mobile phone hardware and the desire by OEMs to offer differentiators have fragmented the market and made portability a far goal to reach, at least in reality.

- The mobile phone hardware was not ready in terms of capabilities or power awareness. This impacted (limited) the way J2ME and BREW were designed.

- The hardware lacked innovative interaction devices. In fact, a QWERTY keyboard upgrade was big news for a handset, a move away from the tedious i^{th} element or *T*9 style keyboard input. This directly limited the UI of most applications developed on these platforms.

- The platforms were highly "regulated" and controlled by the OEM, the operator, or both, leaving the developers' communities with little room for innovations. Some of these regulations were burdening and/or cost prohibitive to the developers, which made it difficult for them to bear any serious revenues out of their applications. BREW, for instance, required costly testing of developed applications on each handset model a developer wished to support. Controlling the platform was practiced even at the API level. For instance, the Location API for J2ME was not made available to many developers, without a prior established relationship with the operator and obtaining an operator-level certificate required for LBS applications.

- Finally, the "*app ecosystem*" was missing. Operators exaggerated in the scrutiny and selectiveness of the developers and their apps. In fact, developers had to first earn the blessings of the operators before they could qualify to getting their apps on the "deck." This posed a huge entry barrier to many developers, which hampered innovation. Application listings were rudimentary to say the least, and the purchase process was very cumbersome to the consumer. Nextel subscribers, for instance, had to remove the back battery cover, then the battery itself, to reach to the SIM card and find the International Mobile Equipment Identity (IMEI) number, required to complete the purchase of J2ME applications. Also there was no centralized billing and application distribution system for J2ME. Developers had to work with different manufacturers, operators and third party distributors to sell applications. The J2ME application signing process also hindered innovation. The Certificate itself cost hundreds of dollars a year. Such cost combined with the then common 50-50 revenue split put most of the investment risk squarely on the developer. The overall experience was mediocre. It was lacking exactly what Steve Jobs had later found necessary for Apple's iPhone to take off—a friction-free App Store. Today, there are many flashlight applications in the App Store (a trivial app that brightens the iPhone screen or the flash LED light to brighten up the user surroundings). There is Flashlight, vFlashlight, Flashlight pro, MacLight, and even Brightest Flashlight Pro! But there is also over half a million applications in the App Store, in addition to *Flashlight*.

1.4 THE STARS ALIGNED

It is as if the past 20 years of tenacious efforts by human technologists to invent the mobile platform have been watched from high above; and what came next was a complete alignment of the stars.

Initial innovations in low-power mobile processors and higher-quality displays demonstrated an exciting and promising future for mobile platforms, especially as gaming and multimedia capable devices. They accomplished a little more; they stimulated the rise of other critical innovations. Better displays, for instance, eased the disappearance of the keyboard, which was finally virtualized. Displays finally came of age and became sophisticated multi-touch screens, powered by intuitive input gestures. Low-power, meter-accuracy GPS technologies (including tower-assisted GPS, or aGPS) also found their way to most handsets, which enabled and fueled location based services. And the design of mobile platforms has gradually relied on the use of embedded sensors to optimize the platform operation (e.g., digital compass, proximity and accelerometer sensors), and to offer higher-level services to the users (e.g., digital cameras and NFC sensors). It is clear now that while we perceived a healthy vision two decades ago, many technological innovations were needed on the hardware and device levels for this vision to pan into reality.

As a consequence to (and along side) the hardware innovations and refinements, new software architectures and operating systems were created from the ground up to take advantage of, and to support, the newest and greatest of mobile device hardware features.

We end our historical coverage here and focus our lecture now on the outcomes and realities of the past several years' shake up in the mobile platform industry. Unthinkable mergers and acquisitions, alliances and marriages, and separations and abandonments, have reshaped the mobile landscape and resulted in three powerful native platforms: Apple's iOS, Google's Android and Microsoft's Windows Phone. Additionally, the Mobile Web has emerged as a device-agnostic platform that can be used from within the native platforms as they embrace the mobile web standards. We cover each platform in a separate chapter, and dedicate a cross-cutting chapter to look at location based services support in each of the platforms. We use a sample application in Appendix A to demonstrate the use and capabilities of each platform. Finally, we speculate a little about the future of mobile platforms, but not in any substantial way that spoils the enjoyment of what we have today.

CHAPTER 2

iOS

iOS is Apple Inc.'s operating system which powers all its mobile devices, ranging from the iPod touch media players to the iPad tablet (`http://www.apple.com/ios`). It was designed from the ground-up as an operating system for touch screen devices and is based on the XNU kernel of Mac OS X. In this chapter, we will cover the history of iOS, a brief primer on multi-touch user interaction, the basic architecture and components of iOS and a look at the software development tools for iOS, and how one can develop their own applications for this wildly popular mobile device platform.

2.1 EVOLUTION: FROM iPHONE OS TO iOS

iOS was originally released in 2007 as the iPhone OS and was the operating system for the first iPhone. It was the first commercial mobile operating system to provide a multi-touch capable user interface and started the industry trend for touch-only mobile devices. The original iPhone OS did not have any support for third-party applications. In 2008, the iPhone OS Software Development Kit (SDK) was introduced for Mac OS X, which allowed developers to create third-party native applications from the iPhone. Subsequently, in the same year, Apple also introduced the App Store where developers could publish their applications and which allowed users to download and install both free and paid applications onto their iPhones. This led to a massive change in the mobile device industry, where the smart phone was no longer seen as a device exclusively used by businessmen or the technically adept. Rather, it became a true mass market device whose capabilities could be changed, customized, and upgraded using third-party applications available through web-based stores. Part of the reason for the App Store's success was that it provided a single storefront to the user for searching and purchasing Apple-certified approved applications instead of forcing the user to go to multiple websites to download and install applications.

In 2010, with the growing success of the iPhone, Apple decided to use the same operating system in its other device offerings and renamed the iPhone OS to iOS. Currently, iOS runs on the iPhone, iPod touch music player, the iPad range of tablets, and the Apple TV set top box. The user interface is consistent across all the devices except Apple TV which has a completely different user interface design. However, iOS does contain specific customizations for each type of device that it runs on. For example, the iPad tablets have support for a split keyboard and tabbed browsing, whereas the other devices do not.

2.2 DIRECT MANIPULATION THROUGH MULTI-TOUCH

One of the most highlighted features of iOS for consumers has always been its multi-touch capability which allows users to use multiple fingers simultaneously to interact with the mobile device using intuitive touch gestures. Multi-touch based interaction is a form of direct manipulation where users directly manipulate digital objects in a physically natural manner and receive continuous feedback on the effect of their input actions. One of the most popular multi-touch gestures on iOS is the "pinch-to-zoom" gesture which can be used to zoom in and zoom out of pictures or other visual content. Other common examples of multi-touch gestures include the use of multiple fingers to rotate an image and using a four-finger swipe up to bring up the task switcher. However, an equally important use of multi-touch, especially on larger devices such as the iPad tablets, is during touch-typing while using the software keyboard where the user can simultaneously tap on multiple keys just like a regular hardware keyboard instead of pressing keys one at a time.

iOS interprets multi-touch gestures as a sequence of single touch inputs. A multi-touch sequence starts when the first finger comes in contact with the touch sensor overlaid on the device display and the sequence ends when the last finger has been removed from the touch sensor. Figure 2.1 shows an example of the "pinch-to-zoom" gesture being used for zooming into a picture. In this gesture, the user places two fingers on the display of the mobile device, depicted by Touch Point #1 and Touch Point #2. The initial position of the two fingers on the display mimics a pinching action. When the user moves one of his fingers relative to the other, the touch sensor detects the motion of the touch points relative to each other. Based on the relative increase in distance between the two touch points as compared to their initial position, the device computes the zoom factor and correspondingly magnifies the picture shown on the display. The result is a very intuitive way of zooming into and magnifying contents displayed on the device screen.

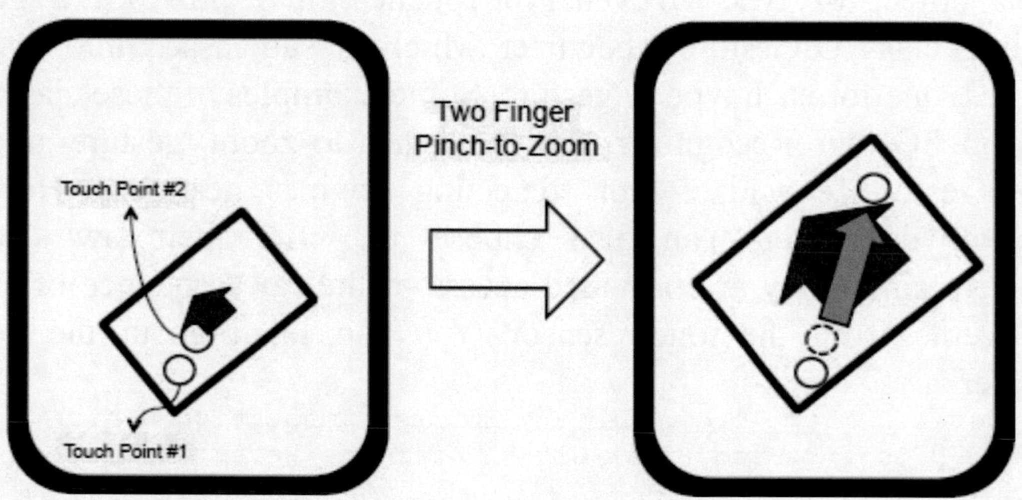

Figure 2.1: Pinch-to-Zoom Multi-touch Gesture.

On the hardware side, all multi-touch devices using iOS use projected capacitive touch sensing technology. This sensor technology typically consists of an insulator, such as glass, which is etched with transparent indium tin oxide (ITO) in the form of a grid. The sensor projects its electrostatic field above the glass so that if a conductor such as a human finger is brought in close proximity to the glass, it distorts the field which is then measured as a change in capacitance. The location of the finger is determined by measuring the change in capacitance at each point on the grid and then transmitted to the operating system's gesture recognition engine. Since the projected capacitive touch technology does not require physical contact between the user's finger and the device display glass, it provides the user with the perception that they only need to touch or swipe very lightly rather than having to physically press down on the display for performing touch-based input. Furthermore, displays with capacitive touch sensor overlays appear brighter and provide better image quality as compared to displays utilizing other technologies, such as resistive touch, since the capacitive touch sensor coatings have lower visible absorbance (that is, absorb less of the light coming from the display). These factors combined together result in a higher quality user experience for the end-user.

On the software side, in iOS a multi-touch gesture is represented as a UIEvent object of type UIEventTypeTouches. iOS provides a gesture recognizer class UIGestureRecognizer which is subclassed into multiple subclasses, one for each type of gesture. Some examples of these classes are the UIPinchGestureRecognizer for the pinch-to-zoom gesture and the UISwipeGestureRecognizer for detecting swipe gestures. However, application developers can also choose to write their own gesture recognizers since they are provided access to the raw sequence of single-touch events from the touch sensor. You can learn about the gesture recognizer here: `http://developer.apple.com/library/IOs/#documentation/UIKit/Reference/UIGestureRecognizer_Class/Reference/Reference.html`

2.3 iOS ARCHITECTURE LAYERS

The iOS software architecture consists of four distinct layers: Core OS Layer, Core Services Layer, Media Layer and Cocoa Touch Layer, as shown in Figure 2.2. Each layer is associated with several frameworks. A framework is a package consisting of a dynamic shared library along with header files defining the application programming interfaces (API) to the library and other resources such as helper applications. Frameworks can be used in an application by simply linking them with the application project. Apart from frameworks, iOS also provides access to standard Unix shared libraries.

2.3.1 CORE OS LAYER

The Core OS Layer is the base layer of iOS which deals with low-level functionalities related to network configuration and communication, accessing the file system, memory allocation, virtual memory management, threading and inter-process communications. The LibSystem library of iOS allows developers to access a number of these functionalities directly. However, for most developers it is much more convenient to utilize frameworks and associated APIs available in the higher layers for their applications. One of the exceptions is the case where the application needs to interact with external hardware such as Bluetooth Low-Energy devices and accessories which are attached to an iOS device through Apple's proprietary 30-pin dock connector. In such cases, the developer needs to utilize the Core Bluetooth or the External Accessory framework, which are part of the Core OS Layer. This layer also provides the Accelerate framework which contains methods for performing image processing and signal processing calculations, optimized for running on Apple's mobile device hardware. Furthermore, the Security framework and the Generic Security Services framework provide security related functionalities to the developer such as managing certificates and trust policies, public and symmetric key cryptography and managing credentials.

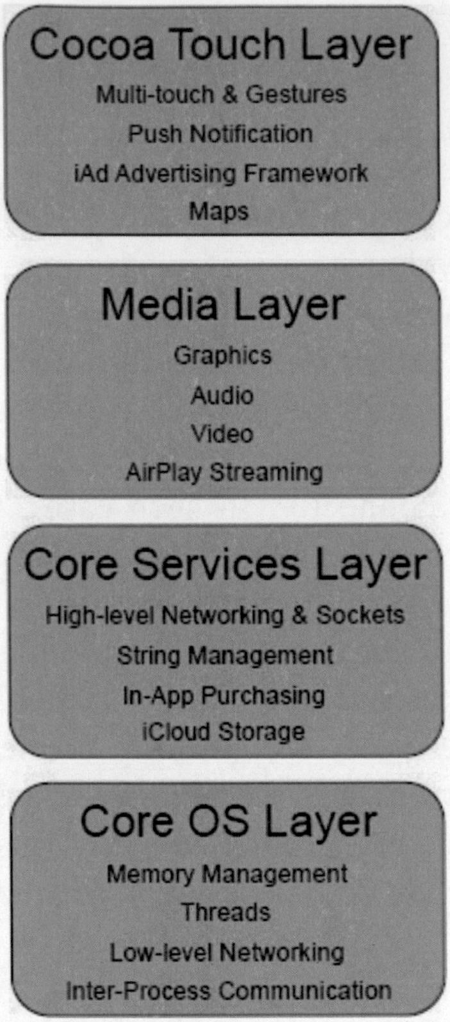

Figure 2.2: The iOS Software Architecture.

More details about the Core OS Layer can be found here: `http://developer.apple.com/library/ios/#documentation/Miscellaneous/Conceptual/iPhoneOSTechOverview/CoreOSLayer/CoreOSLayer.html`

2.3.2 CORE SERVICES LAYER

The Core Services Layer provides access to the key system services that are required by applications. One of the main frameworks in this layer is the Core Foundation framework, which provides capabilities such a string management, threading, socket communication and managing data types such as arrays, URLs and even raw bytes. The Core Foundation framework

APIs are C-based. However, the Foundation framework provides Objective-C wrappers for the interfaces provided by the Core Foundation framework. The CFNetwork framework provides networking capabilities using BSD style sockets and also provides higher-level functionality for communication using common protocols such as Hyper Text Transport Protocol (HTTP) and File Transfer Protocol (FTP), including secure communications using Secure Socket Layer (SSL) and Transport Layer Security (TLS).

A major addition to the Core Services layer has been the support for iCloud Storage, which was introduced in iOS5 in 2011. This allows third-party applications to store data and access them from any of the user's iOS devices in a simple seamless manner. iCloud provides two kinds of data storage: one for storing documents and data in the user's iCloud account and the other for storing data which can be shared between multiple instances of the same application across different iOS devices. We will discuss more about the iCloud in Section 2.5.

Another fundamental capability that is available through the Core Services Layer is the In-App Purchases, which enables developers to sell content and services and perform iTunes financial transactions from within their application. This capability is available through the Store Kit framework, which handles the transactional and security aspects of in-application purchasing.

Apart from these, the Core Services Layer contains a large number of frameworks for enabling different application capabilities. The Address Book framework provides access to a user's phone book and contacts including the capability to modify them. The Accounts framework enables applications to utilize the single sign-on model so that they do not need to unnecessarily prompt the user for login information. The Core Location framework provides location information to applications for creating location-based services (LBS). It utilizes multiple technologies such as WiFi localization, cell tower positioning and on-board GPS to determine the device's location and provides that information to the application.

Apart from frameworks, this layer also provides standard shared libraries such as the SQLite library for embedding SQL databases within applications and an XML library based on libXML2 for standard XML parsing and manipulation.

Learn more about the capabilities of the Core Service Layer at: http://developer.apple.com/library/ios/#documentation/Miscellaneous/Conceptual/iPhoneOSTechOverview/CoreServicesLayer/CoreServicesLayer.html

2.3.3 MEDIA LAYER

One of the main strengths of the iOS user experience has always been its rich multimedia capability. The Media Layer is responsible for enabling developers to write applications centered on rich user interaction by utilizing the high quality audio-visual capabilities of an iOS device. The Media Layer's capabilities can be classified into four different categories: Graphics, Audio, Video and Streaming using AirPlay.

Mobile devices from Apple have always had an emphasis on high-quality high-resolution displays such as the Retina display and hence, the graphics capabilities provided by iOS are quite powerful. The Core Graphics framework provides interfaces for basic 2D drawing and rendering capabilities using Quartz, which is a vector-based drawing engine originating from OS X. The APIs are C-based and cover functionalities ranging from coordinate transformations, to anti-aliasing and path-based drawing. In addition, the Quartz Core framework provides capabilities for animating an application's user interface while leveraging hardware graphics acceleration. These are probably one of the most visible aspects of iOS applications from a user's perspective. This framework also provides Objective-C wrappers to its APIs. The Open GL ES framework enables developers to create and manipulate 2D and 3D content in their applications especially for games. OpenGL ES (or, OpenGL for Embedded Systems) is the embedded devices version of the popular OpenGL specification for rendering 2D and 3D graphics. Starting with iOS5, the GLKit framework provides a number of useful classes which simplify the creation of an OpenGL ES application by abstracting away some of the low level details. The Core Text framework provides advanced text layout and rendering capabilities beyond what is provided by the UIKit framework. For image manipulation and format conversion, the Media Layer provides the Core

Image framework and the Image I/O framework. These frameworks utilize the Graphics Processing Unit (GPU) hardware of the iOS device and can even be used for advanced tasks such as face detection. Apart from the frameworks, which provide multimedia handling capabilities, the Media Layer also contains an Assets Library framework which allows a developer to retrieve photos and videos from iOS devices including pictures shot with the device's in-built camera.

The main Media Layer framework for audio- and video-related functionality is the Media Player framework. This framework not only supports the playing of music tracks and videos but also allows access to the iTunes music library of the user and playlist management. The lower-level functionality of audio recording, mixing and playback is actually handled by the Core Audio framework. However, if developers are interested in heavy utilization of audio and video recording and playback in their applications, they must use the AV Foundation framework. The AV Foundation framework sits between the Core Media/Audio and the Media Player framework. It provides Objective-C wrappers for audio manipulating but now also offers functions for capturing and editing movies and streaming audio and video using Apple's AirPlay technology, which was introduced in 2010. AirPlay allows a user to stream both audio and video wirelessly to another iOS devices. The content is either mirrored to what is being displayed on the originating device or specific content is routed to specific devices based on the application developer's choice. Finally, the OpenAL framework provides the capability to utilize positional audio in application through the OpenAL standard.

More details about the Media Layer can be accessed here: `http://developer.apple.com/library/ios/#documentation/Miscellaneous/Conceptual/iPhoneOSTechOverview/MediaLayer/MediaLayer.html`

2.3.4 COCOA TOUCH LAYER

The Cocoa Touch Layer is responsible for the user interaction capabilities that can be utilized to create iOS applications. It provides access to the multi-touch input capabilities that have made iOS devices so popular. However, it also provides access to other important features such as multitasking, in-application advertising, push notifications, maps, wireless printing, file sharing, peer-to-peer Bluetooth connectivity and support for external displays.

The main framework in the Cocoa Touch Layer is the UIKit framework which provides basic interaction capabilities for iOS applications. This includes capabilities such as touch interaction and multi-touch gesture recognition, creation of user interfaces, multitasking support, wireless printing and connecting to external displays through wired interfaces. The UIGestureRecognizer class, which was described in Section 2.2 earlier is a part of the UIKit framework and used for detecting common multi-touch gestures such as "pinch-to-zoom" and swipes. The UIKit framework also provides access to sensors embedded in iOS devices such as accelerometer, camera, proximity sensor and battery monitor. The multitasking functionality provided by iOS is slightly different from the traditional definition. iOS does not support true multitasking where an application keeps running even when sent to the background. Instead, when an application is sent to the background it is suspended either immediately or within a finite amount of time, unless it is utilizing specific iOS services which are allowed to run in the background. When the application is brought to the foreground and is shown on the device display then it once again resumes execution. A related feature is the Push Notification, which enables applications to alert users using text or audio cues even when the application is not actively running.

The iAd framework was introduced in iOS4 and lets developers display advertisement banners inside their application. This framework abstracts away the low-level functionalities required for loading and presenting ads and responding to user clicks. This provides an extremely powerful way for developers to further monetize their applications apart from the usual revenue generated through downloads of their applications from the Apple App Store.

The Map Kit framework makes it really easy to insert a map interface into an application. The map can then be augmented with directions, points of interest and other application specific information including customized images and multiple layered overlays. Navigation through the map can either rely on user input or can be done programmatically.

There are several other frameworks in the Cocoa Touch Layer, such as the Message UI framework for composing and handling email messages, the Address Book UI framework for creating, editing and displaying contacts and the Game Kit framework for adding peer-to-peer communication in applications. Another interesting and recently introduced framework is the Twitter framework which came out with iOS5. It provides support for the popular micro-blogging service, Twitter, and handles the creation and sending of tweets from within any iOS application in a simple and easy manner.

Interested readers can learn more about the Cocoa Touch Layer at: `http://developer.apple.com/library/ios/#documentation/Miscellaneous/Conceptual/iPhoneOSTechOverview/iPhoneOSTechnologies/iPhoneOSTechnologies.html`

2.4 SOFTWARE DEVELOPMENT TOOLS

The iOS Software Development Kit (SDK) was released in 2008 and enables third-party developers to create and deploy applications to iPhone, iPad and iPod touch devices. The iOS SDK was designed so that it would not only allow application developers already familiar with the Mac OS X platform to migrate their code to run on iOS mobile devices but also provide a low barrier for novice developers to create rich interactive applications. You need to register and create a developer account with the iOS Dev Center at: `http://developer.apple.com/devcenter/ios/` and have an Intel-based Mac computer running Mac OS X Snow Leopard or later to install the SDK. Currently, the SDK does not work on Windows or Linux systems.

2.4.1 OBJECTIVE C

Objective C is an object-oriented programming language, which is used for developing applications for iOS. It is essentially an additional object-oriented layer with Smalltalk-style messaging on top of the standard C language. Hence, regular C programs can be compiled using an Objective C compiler and many frameworks within iOS also provide C style interfaces. Objective C works on the models of message passing where instead of method calls, messages are passed to the target method. Objective C also provides run-time support for reflection which allows a program to query and modify its own internal structure and hence, behavior during execution. An important thing to note about Objective C especially for C++ programmers is that Objective C does not support operator overloading and also does not allow multiple inheritance.

2.4.2 XCODE

Xcode is the Integrated Developer Environment (IDE) Apple for writing applications for iOS and Mac OS X. Xcode comes with a suite of developer tools and comprehensive documentation. It utilizes a modified version of the GNU Compiler Collection (GCC) for compiling Objective C programs and uses the GNU Debugger (GDB) back-end for debugging. Xcode compiles iOS applications to run on ARM-based processors which are used in all the iOS devices. One of the interesting features of Xcode is that it allows the distribution of compiling and building applications across multiple networked computers, a capability known as Dedicated Network Build.

2.4.3 INTERFACE BUILDER

The Interface Builder is part of the Xcode suite of tools and enables developers to create application user interfaces using GUI based interaction. Interface Builder provides the developer with groups of user interface elements (called palettes). Developers can visually design their application's user interface and also specify the connections between the UI elements and the application code.

2.4.4 INSTRUMENTS

Instruments is a tool used for profiling and analyzing the performance of an iOS application. It uses the DTrace tracing framework. It can be used to not only track CPU load, memory allocation events, graphics rendering calls which use OpenGL ES, file activity and networking overheard, but also track user input events occurring during the runtime of the application. This is an essential tool for the developer to ensure that their iOS application is completely optimized for smooth user interaction and efficient in its use of system resources.

2.4.5 iOS SIMULATOR

The iOS simulator provides the ability to run an iOS application on a Mac computer without requiring the application to be deployed on an actual iOS device. Currently, the iOS simulator can simulate iPhone and iPad devices running different versions of iOS. The advantage of this tool is that the developer can do rapid iterative testing and modification of their application logic and UI before deploying it to an actual device. It also provides a way to ensure that most of the fundamental bugs in the application are weeded out before it is actually deployed and run for the first time on an iOS device. The iOS simulator provides the capability to simulate specific actions such as rotating and shaking the simulated device and simulating a TV out connection. However, it does not have the capability to simulate the readings from the accelerometer sensor or images captured from the camera.

2.4.6 WRITING YOUR FIRST iOS APP

The best way to learn how to develop application for iOS is to go through the tutorials provided with Xcode and try to run and modify the sample applications. If you are still deciding whether iOS is the right choice of platform for you, you can use the iOS simulator to try out different applications for free. You can also refer to the example TwitterSearch application and source code provided in Appendix A.1.

2.5 iCLOUD

iCloud is Apple's cloud storage service where applications can store data in remote servers (referred to as "the cloud") instead of locally on the device. This provides a two-fold advantage: not only can users and applications access that data across multiple devices, but even if a user loses an iOS device, the data is not lost and is still securely stored in the cloud and hence can be quickly retrieved over the network and used on another iOS device. However, it is important to note that iCloud is not intended to support multi-user access or behave in any way like a distributed file system. Rather, it is intended to provide a loosely (yet seamlessly) synchronized storage across multiple devices owned by a single user. Only "eventual consistency" among the devices is guaranteed by iCloud, which, for a single user, is more than adequate and acceptable.

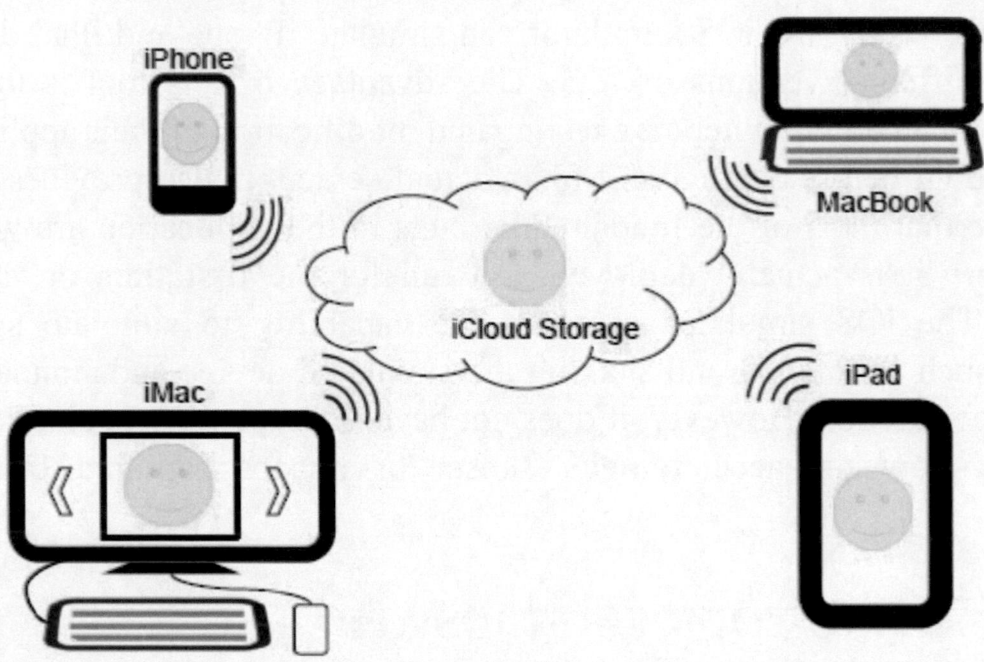

Figure 2.3: iCloud Remote Storage across Multiple Devices.

The iCloud service provides two categories of storage for applications: iCloud Document Storage and iCloud Key-Value Data Storage.

The iCloud document storage allows applications to store documents and data in the user's iCloud account. This is the most commonly used functionality for most applications especially for storing user created content in the cloud. It is recommended that relatively critical data which needs to be stored in the cloud must be done using this type of storage. Any changes made to the stored content on one device are automatically pushed to the cloud and other devices are notified of the change. The frequency of updates pushed to the cloud back-end are optimized and handled internally by iOS and do not require implementation of any additional logic in the application. In response to an update, an application can trigger a download of the changed content by either explicitly downloading it or by attempting to access the content file which causes the Core Services Layer to automatically download and synchronize the local copy with the updated content. It is important to understand that iCloud resources must be used wisely by the developer, and content stored in the iCloud storage must be of the type which cannot be generated or replicated by the application.

The iCloud key-value storage provides a mechanism for applications to share up to a few KB of application-specific configuration information between multiple instances of the application running across different iOS devices. It is basically a hash table of application-specific properties which is periodically synchronized between the iOS device and the cloud back-end. Storage is limited to 64 KB per application and is also equivalent to the maximum size of data that can be stored for one key.

The iCloud service allows iOS developers to write applications which cannot only ensure better data security and accessibility but also provide better user experience by enabling seamless transition and synchronization of documents, media and user preferences across multiple devices. You can learn more about the exciting opportunities made available through iCloud at: `http://www.apple.com/icloud/what-is.html`

Table 2.1: Some Useful iOS References

Apple iOS Website	`http://www.apple.com/ios`
iOS Dev Center (Developer account, SDK download)	`http://developer.apple.com/devcenter/ios/`
Apple iCloud	`http://www.apple.com/icloud/what-is.html`

Table 2.2: References for Some Useful iOS Components

Function	Component	Reference URL
Low-level System Functions	Core OS Layer	http://developer.apple.com/library/ios/#documentation/Miscellaneous/Conceptual/iPhoneOSTechOverview/CoreOSLayer/CoreOSLayer.html
High-level System Interfaces, In-App Purchasing, iCloud Remote Storage	Core Services Layer	http://developer.apple.com/library/ios/#documentation/Miscellaneous/Conceptual/iPhoneOSTechOverview/CoreServicesLayer/CoreServicesLayer.html
Audio, Video, 2D and 3D Graphics, AirPlay Streaming	Media Layer	http://developer.apple.com/library/ios/#documentation/Miscellaneous/Conceptual/iPhoneOSTechOverview/MediaLayer/MediaLayer.html
Multi-touch Interaction, Gesture Recognition, iAd advertising, Maps and Location-based Services	Cocoa Touch Layer	http://developer.apple.com/library/ios/#documentation/Miscellaneous/Conceptual/iPhoneOSTechOverview/iPhoneOSTechnologies/iPhoneOSTechnologies.html

CHAPTER 3

Android

The Android platform is developed and maintained by the Open Handset Alliance which is led by Google (`http://www.openhandsetalliance.com`). The Open Handset Alliance consists of many major mobile device manufacturers and chipset vendors such as Samsung, Motorola, HTC, Intel, Qualcomm and Texas Instruments. Apart from these, a number of mobile network operators and software companies are also part of the consortium. It was founded in 2007 and has seen a rapid growth in its membership from the original 37 members to 84 members today.

Android is available as an open source mobile platform under the GNU General Public License and the Apache 2.0 license. This enables anybody who is interested, to download and customize its source code for their own use without paying any licensing fees or royalty to the consortium. However, any changes to the core platform are tightly controlled by the Open Handset Alliance and need to be approved by Google. Moreover, the applications, which are provided by Google such as Maps and Gmail, are closed source and only provided to those whose devices pass a specific certification process. Each major update of the Android platform is identified by the name of a desert and the current version is called Ice Cream Sandwich and has the version number 4.0.

3.1 FROM HUMBLE BEGINNINGS TO TOP DOG

Android had its early beginnings as the flagship product of Android Inc., a small startup founded in 1999 by Andy Rubin along with Chris White, Nick Sears and Rich Miner to develop a new user-friendly location-aware mobile platform. After surviving through a rocky start, Android Inc. was acquired by Google in 2005. While the platform was being further developed in-house at Google, a number of hardware, handset and network partners were also solicited for partnership and in 2007 the Open Handset Alliance was formally announced. The first Android phone running Android version 1.0 was launched in September 2008 branded as the HTC Dream. However, with a rapid sequence of updates and a growing number of partners launching their own mobile devices, including highly popular devices such as the Motorola Droid and the Samsung Galaxy S family of phones, Android experienced explosive growth and according to some analysts has already overtaken Symbian as the dominant smartphone platform in the world with no signs of slowing down any time soon. This coupled with the fact that apart from phones Android is also used across a wide variety of mobile devices such as tablets, personal music players and even netbooks and in-flight entertainment systems, makes it a very attractive mobile platform from a developer's perspective. Currently, it is estimated that Android runs on over 250 million devices around the world and is still growing at an unparalleled rate.

3.2 PLATFORM ARCHITECTURE

Android is a Linux-based platform and uses a heavily customized Linux kernel, however it cannot be considered as just another flavor of Linux given the fact that it does not support the complete set of standard GNU libraries and uses its own proprietary windowing system instead of X-Windows. Here, we will take a look at the various system components, which make up the core platform. Figure 3.1 shows an overview of the Android platform architecture.

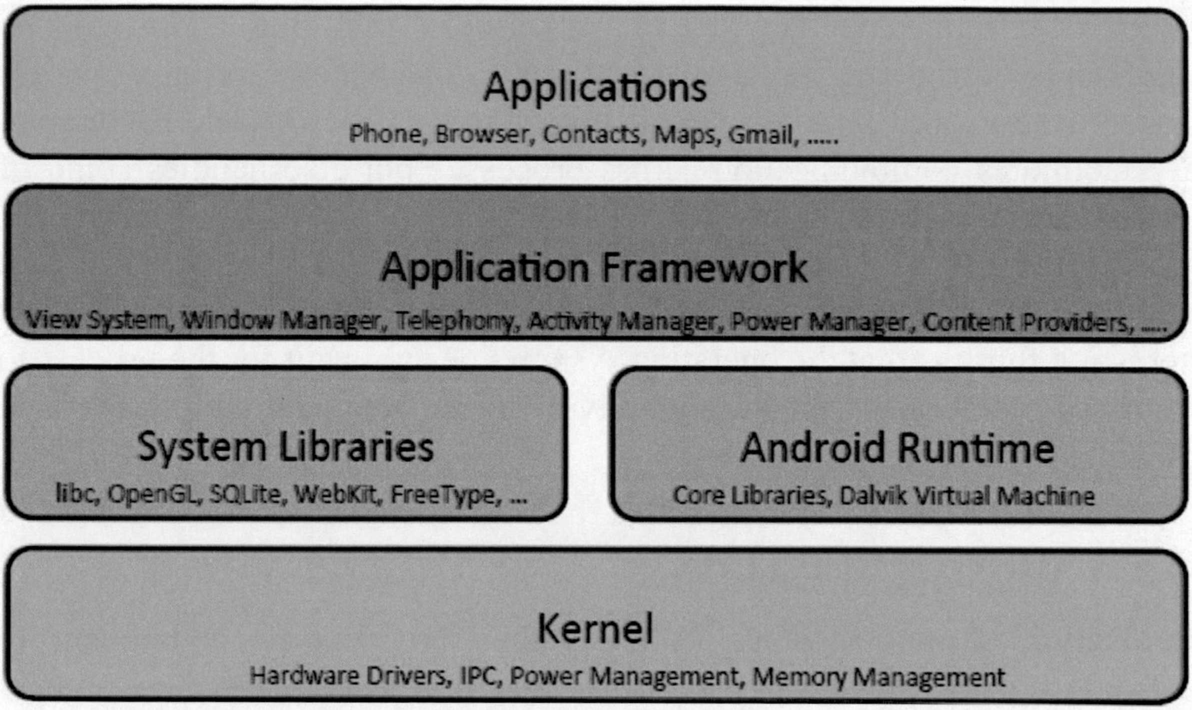

Applications
Phone, Browser, Contacts, Maps, Gmail, ...

Application Framework
View System, Window Manager, Telephony, Activity Manager, Power Manager, Content Providers, ...

System Libraries
libc, OpenGL, SQLite, WebKit, FreeType, ...

Android Runtime
Core Libraries, Dalvik Virtual Machine

Kernel
Hardware Drivers, IPC, Power Management, Memory Management

Figure 3.1: Android Platform Architecture.

3.2.1 KERNEL

The Android kernel is currently based on version 2.6 of the Linux kernel. The stock Linux kernel was customized and several patches were applied to overcome shortcomings or introduce new functionality. The kernel contains the hardware abstraction layer (HAL) and provides drivers for the display, touch input, networking, power management and storage. It also contains components for memory management and inter-process communication amongst other low-level functionalities. One of the main features added was YAFFS2 (Yet Another Flash File System 2) which enables support for flash-based file systems. This is considered especially important for mobile devices which almost universally use solid state flash-based storage instead of conventional hard drives. Another important yet controversial feature introduced in the Android kernel is the concept of WakeLocks. A WakeLock can be used to force a device from going into a low power state and is useful from a user experience perspective since it makes the device more responsive to user interaction. It is primarily used by applications and services to request CPU resources essentially implying that if the operating system does not have any active WakeLocks then the CPU will be shutdown. Android utilizes its own proprietary mechanism for inter-process communication and remote method invocation called Binder. Binder not only facilitates communication across processes but also handles some of the associated memory management and management of the lifecycle of objects which are shared across processes. Android also provides the capability to kill processes which are running low on memory. The interesting thing is that the operating system does not wait for the process to run out of memory but rather aggressively terminates them before they get into an out-of-memory situation.

3.2.2 ANDROID RUNTIME

The Dalvik Virtual Machine (VM) is the core runtime component of Android. It is a process-based virtual machine which uses the register architecture and is optimized for low memory footprint and better performance efficiency. Dalvik runs on top of the Android kernel and uses it for low-level functions such as multi-threading and memory management. Applications in Android are written in a dialect of Java which is compiled into bytecode and stored as Java Virtual Machine (JVM) compatible *.class* files. However, there are major differences between standard Java APIs and the Android APIs and the JVM is not used for executing Android applications. A major difference is the absence of Abstract Window Toolkit (AWT) and Swing libraries in Android. The JVM compatible class files are converted into Dalvik Executable *(.dex)* files which are then executed by the Dalvik VM when the application is run on an Android device. The current version of Dalvik also includes a Just-In-Time (JIT) compiler which improves run-time performance. Each Android application executes within its own instance of the Dalvik VM which is in turn run as a kernel managed process. The Android operating system is designed to ensure that multiple instances of the VM can run at the same time without affecting user experience. The sandboxed approach ensures security and if an application requires access to device data outside its own sandbox, such as contacts, text messages and Bluetooth communications, it needs to explicitly request permission during installation on a specific device.

3.2.3 SYSTEM LIBRARIES

Android provides a number of system libraries in C/C++ which are made accessible through the Application Framework. It must be noted that Android does not provide the complete functionality required of the standard GNU C libraries in Linux. The most common library provided by Android which should be familiar to C/C++ developers is libc. The FreeType library provides bitmap and vector-based font rendering. The SQLite library enables developers to add database capabilities into their applications. Android supports OpenGL/ES 1.0, 1.1 and 2.0 APIs for 2D and 3D rendering which can utilize hardware acceleration if available on the device. The Scalable Graphics Library (SGL) is responsible for all the 2D rendering tasks. The compositing of 2D and 3D content is handled by the Surface Manager library.

One of the key features of the Android platform is its web-driven capabilities. The LibWeb-Core library provides a WebKit-based browser engine which can be embedded as a web view within user interfaces of other applications. The stock Android browser application is also built using this library. Moving forward, the web browser library will prove to be extremely important for application developers with the rise of HTML5 web applications and cloud computing where applications will be developed for and executed within a browser as opposed to a native operating system and all the data will be stored in a remote cloud back-end instead of locally on the device. The Android Media Library is based on the OpenCORE multimedia framework developed by PacketVideo. This library handles both images and multimedia content and provides capabilities for recording and playback of commonly used audio, video and still-image formats such as MP3, AAC, H.264, MP4, JPG and PNG.

3.2.4 APPLICATION FRAMEWORK

The Android Application Framework provides a high level API for application developers to take advantage of the various capabilities of the platform. The View System provides the basic building blocks for creating the user interface of an application. A View object in Android represents a rectangular region of the device's display and is responsible for all the rendering and event handling tasks within that region. Visual user interface elements in Android are called Widgets and are derived from the View class. Developers can also create their own custom widgets. The ViewGroup subclass is the base class for layouts which are essentially containers which encapsulate other View objects and determine their on-screen layout in terms of position, spacing and orientation. All views within the same window are arranged hierarchically in a tree structure and are stored in XML format. Apart from handling drawing, animation and input events, Android views also provide some security capability by allowing applications to request the framework to filter out touch events based on specific contexts. The reader can browse through a rich collection of common views and widgets here: `http://developer.android.com/resources/tutorials/views/index.html`.

The ContentProvider component allows applications to access and share data with other applications. The ContentProvider class allows an application to publish its data for access by other applications. The information is provided through a single ContentResolver interface. When a request for data is made through this interface, it checks its validity and passes it on to the appropriate ContentProvider. The lower-level details of this inter-process communication are abstracted away from the application developer and handled by the operating system. For details on how to create content providers, manage data and query them please visit: `http://developer.android.com/guide/topics/providers/content-providers.html`.

One of the key strengths of the Android user interface has been its notification management system through which any application can notify the user about specific events when they occur. This capability is handled by the NotificationManager class, which allows an application to issue notifications in the form of an icon on the status bar, flashing the device LEDs or the backlight of the device display or by playing a sound or vibrating the device. Optionally, if a user clicks on the icon of a notification, application-specific actions can be executed. Details on the Notification-Manager class can be found at: `http://developer.android.com/reference/android/app/NotificationManager.html`.

The Android Resource Manager provides a way to separate static resources from the application code. These resources can be animations, layouts, strings and even image files. A developer can utilize this capability to customize the application based on the type of device it is deployed in or various device settings, such as locale information and language. Any resource is accessible within the code by addressing it using its package name, resource type and resource name in the following syntax: <package_name>.R.<resource_type>.<resource_name>, where resource_type is set from a pre-determined list of values specified by Android. Utilizing resources also helps the developers in maintaining and updating their applications in an efficient manner. You can learn about how to manage and access resources in an Android application including run-time configuration changes here: `http://developer.android.com/guide/topics/resources/index.html`.

A major selling point for the Android platform has always been its mapping capabilities bolstered by the availability of free high-quality turn-by-turn navigation and map applications such as Google Maps. The LocationManager class provides access to the locationing services available on the device. Third-party developers can utilize this class to create location-based services (LBS) where the application is not only aware of the user's current geographical location but can also ask the operating system to trigger specific actions whenever the device is within the proximity of a specific location. Android also provides specific classes such as GpsSatellite and GpsStatus to access the underlying GPS engine which determines device location from the Global Positioning System (GPS) using the NMEA protocol. More details about the LocationManager class can be found here: `http://developer.android.com/reference/android/location/LocationManager.html`.

The InputMethodService class enables developers to implement their own custom software keyboards, keypads and even pen input. The input is then converted into text and passed on to the target UI element. Learn more about this class at: `http://developer.android.com/reference/android/inputmethodservice/InputMethodService.html`.

The TelephonyManager class provides applications with the ability to determine telephony services on the devices and access specific subscriber information. Applications can also subscribe to telephony state changes. Android supports both GSM and CDMA cellular technologies and applications can access information specific to these technologies depending on the device. The SmsManager allows applications to send data and text messages using the Short Message Service (SMS) protocol. You can learn more about the TelephonyManager at: `http://developer.android.com/reference/android/telephony/TelephonyManager.html`.

A number of utility classes are also provided such as PhoneNumberUtils and PhoneNumberFormattingTextWatcher which simplify the handling and formatting of phone number strings from around the world.

The PowerManager class provides applications with the ability to control the power state of the device and use a feature called WakeLocks (defined by the PowerManager.WakeLock class) which forces a device to remain on and not go into power-saving mode. This feature if used wisely can provide superior user experience by ensuring that the application's interface is immediately responsive to user actions. However, misuse of this capability can lead to extremely poor battery life on the device and hence, developers should carefully assess their requirements and code their application to ensure that WakeLocks are acquired only when absolutely necessary and are released after the shortest possible length of time. You can refer to this link for more details: `http://developer.android.com/reference/android/os/PowerManager.html`.

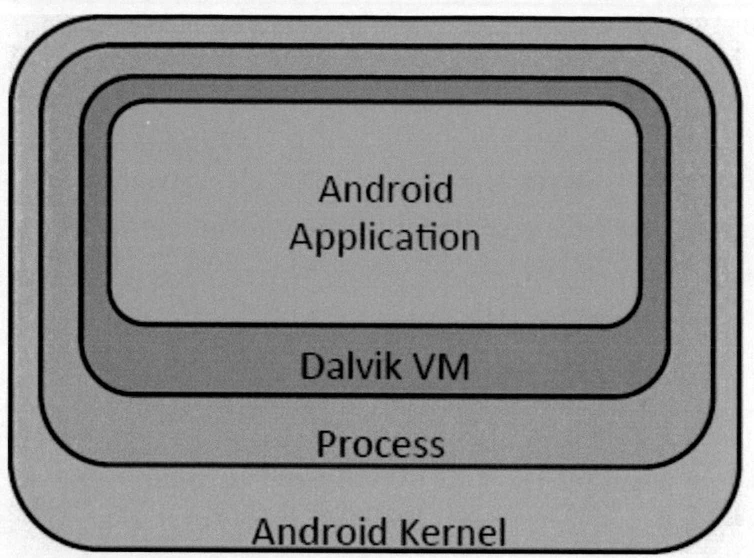

Figure 3.2: Android Application Runtime.

3.3 DEVELOPING ANDROID APPLICATIONS

Android applications are written in Java, however they are executed by the Dalvik Virtual Machine (VM). Even though Java is used as a programming language, Android is not fully compliant with standard Java due to major differences especially in the user interface libraries. Each application runs in a sandboxed environment with its own instance of a Dalvik VM which, in turn, runs within its own kernel managed process. An Android application is installed as a single file of type Android Package (extension: *.apk*) which contains the compiled code along with data and resource files. Android package files can be signed or unsigned.

3.4 ANATOMY OF AN ANDROID APPLICATION

An Android application can be composed of 4 types of components namely: Activity, Service, Content Provider and Broadcast Receiver.

The Activity component represents a screen with a visual user interface and is defined by the Activity class. If an application has multiple user interface screens then it will contain multiple activity components. For example, a music player application user interface consists of one screen for audio playback and one screen for selecting albums or audio files will have two activity components. An interesting capability of Android is that any application can ask the Application Framework to start a specific activity in another application. Hence, if an email application receives a message with an MP3 audio attachment it can request the framework to start the activity representing the music player's audio playback screen so that the user can listen to the contents of the attachment. This is done asynchronously through the use of Intents. An Intent object provides an abstract description of an action be performed in the form of launching an Activity component or a specific type of component.

In contrast, the Service component (defined by the Service class) is used for background tasks such as time intensive tasks or inter-application functionalities which do not require direct user interaction. A service component does not have a user interface. In our music player example if we had the capability to play music in the background, that would be delegated to a service component. A service component can also be started based on a request from another application.

The Content Provider component enables applications to store and share data with other applications. It is defined by the ContentProvider class and enables sharing using the ContentResolver interface. Through this interface other applications can query and if allowed, modify another application's data. In the music player example, the application can share information such as the name of the audio file currently being played with other applications, such as a music store application which can then supply album recommendations to the user for online purchase. It is important to note that the content provider component cannot be activated on request from another application using Intent objects and all the interaction is done through the ContentResolver interface.

The Broadcast Receiver component is defined by the BroadcastReceiver class and is tasked with responding to system-wide broadcast announcements. Android can broadcast a number of system status messages such as device battery status or when the camera has just captured a picture. Each broadcast is represented as an Intent object however it only contains the message associated with the broadcast and does not specify any Activity object that has to be triggered. An application can then define ways to handle specific events based on the message received. In terms of user interaction broadcast receivers can be classified as being somewhere between the Activity and Content Provider components. A broadcast receiver does not have a user interface, however it can create a notification on the device's user interface using the Notification Manager.

All components present in an application must be declared in a manifest file named AndroidManifest.xml. For each component, the manifest file can also declare an Intent Filter to list its capabilities so that it can respond to intents of specific types. Apart from this, the manifest file also declares the user permissions required by the application during run time such as access to user data or network access. It also specifies the hardware and software services required and the external libraries that need to be linked with the application. Since Android has been evolving rapidly with four major versions launched within the last three years, the manifest file also specifies the API Level used by the application. The value of the API level corresponds to a release version of the Android platform. For example, the API level associated with the Ice Cream Sandwich release of Android (version 4.0) is 14.

3.5 THE ANDROID SOFTWARE DEVELOPMENT KIT

Developers interested in writing applications for Android can download the Android Software Development Kit (SDK) from `http://developer.android.com/sdk`. Currently, the SDK is supported on Microsoft Windows, Mac OS X and Linux operating systems. A recommended Integrated Development Environment (IDE) for Android development is Eclipse with the Android Development Tools (ADT) plugin. Eclipse can be downloaded from `http://www.eclipse.org/downloads/`. The ADT plugin extends the capabilities of the Eclipse IDE and provides an easy way to create Android projects, application UIs, debugging and testing. Apart from the basic SDK, developers can also install add-ons such as offline API documentation, code samples including sample components for multiple API levels and APIs for integrating Google Maps based functionality into applications.

3.5.1 DEBUGGING AND TESTING

For debugging applications, developers can use any Java Debug Wire Protocol (JDWP) compliant debugger. However, if Eclipse is being used as an IDE then it already comes with its own JDWP compliant debugger which should work satisfactorily for most developers. The SDK comes with the Android Debug Bridge (adb) tool which can be used for both on-device debugging and for debugging applications running on the software device emulator using an Android Virtual Device configuration. The Android Debug Bridge facilitates communication between the device or emulator executing an application and the development machine and is also used for installing the application on the target device. The Dalvik Debug Monitor Server (DDMS) runs on the development machine and connects to the Android Debug Bridge. It also provides a graphical interface which shows information about running threads and the call stack and also assists in the debugging process by allowing screenshot capture and spoofing incoming calls and SMS events.

The Android Testing Framework is based on JUnit, the Java unit-testing framework. The easiest way to test Android applications is to use the ADT Eclipse plugin, which is capable of creating and building test packages inside Eclipse. The Android Testing API includes an instrumentation framework, which provides hooks, which enable direct access to system methods which control a component's lifecycle. This allows the developer to control the lifecycle of a specific component independently step-by-step in an isolated manner. In terms of application features which must be tested, Google recommends that developers must ensure that their application reacts as expected during change in device orientation, change in device configuration such as hardware capability or locale, unavailability of specific external resources used by the application and the impact on battery life. The impact on battery life must be especially tested thoroughly, given that the Android kernel provides the WakeLock functionality which if misused can severely deplete the device battery.

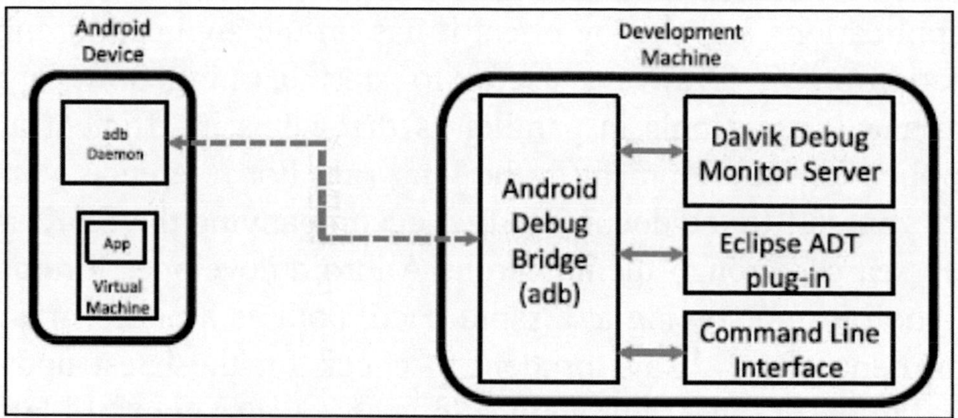

Figure 3.3: Debugging Android Applications.

3.5.2 USEFUL TOOLS

The Android SDK comes with a number of useful tools which can greatly speed up development time and ensure good quality applications. The Android Emulator is a QEMU-based software emulator for the ARM processor architecture. It can be used to emulate various types of devices and associated hardware configurations through the use of Android Virtual Device configurations. The Traceview tool generates graphical views from application execution logs whereas dmtracedump generates call stack diagrams from trace log files. For user interface related development, the Hierarchy Viewer tool provides a visualization of a layout's view hierarchy and has a feature called Pixel Perfect View for inspecting the UI details using a magnifier. For optimization of UI layouts, the layoutopt tool can prove quite useful. It analyzes the layout XML files using a set of pre-defined rules and warns of any rule violations and offers possible solutions.

3.5.3 WRITING YOUR FIRST ANDROID APPLICATION

The fastest way to get your hands dirty with Android development is to install the code samples accompanying the SDK and try compiling and running them both on the emulator and on the real device. You are also recommended to try out the TwitterSearch example application provided with this lecture. Once you are comfortable with the creation, building and deployment of an Android application, you should try modifying pieces of existing applications and see the effect it has on the application during run time. In addition, it is always useful to start learning how to use the debugging and testing tools in parallel as that helps in efficient and high quality application development in the long run. For reference, you should not restrict yourself to the documentation accompanying the SDK and must proactively search through the numerous Android developer forums on the Internet. Android is evolving at a rapid pace, both as a platform and as an ecosystem, hence it is always prudent to check for the latest updates and fixes online to ensure that your applications are always compatible with the latest devices and provide the best user experience possible. Last but not least, as you start exploring Android further, you should always keep the following link to the Android API reference handy: `http://developer.android.com/reference/packages.html`.

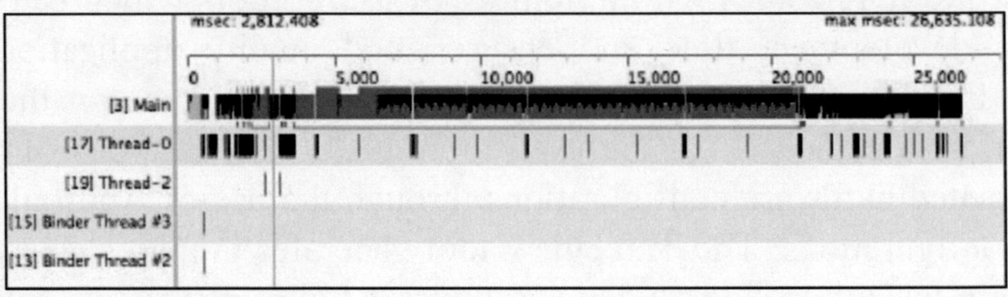

Figure 3.4: The Traceview Tool.

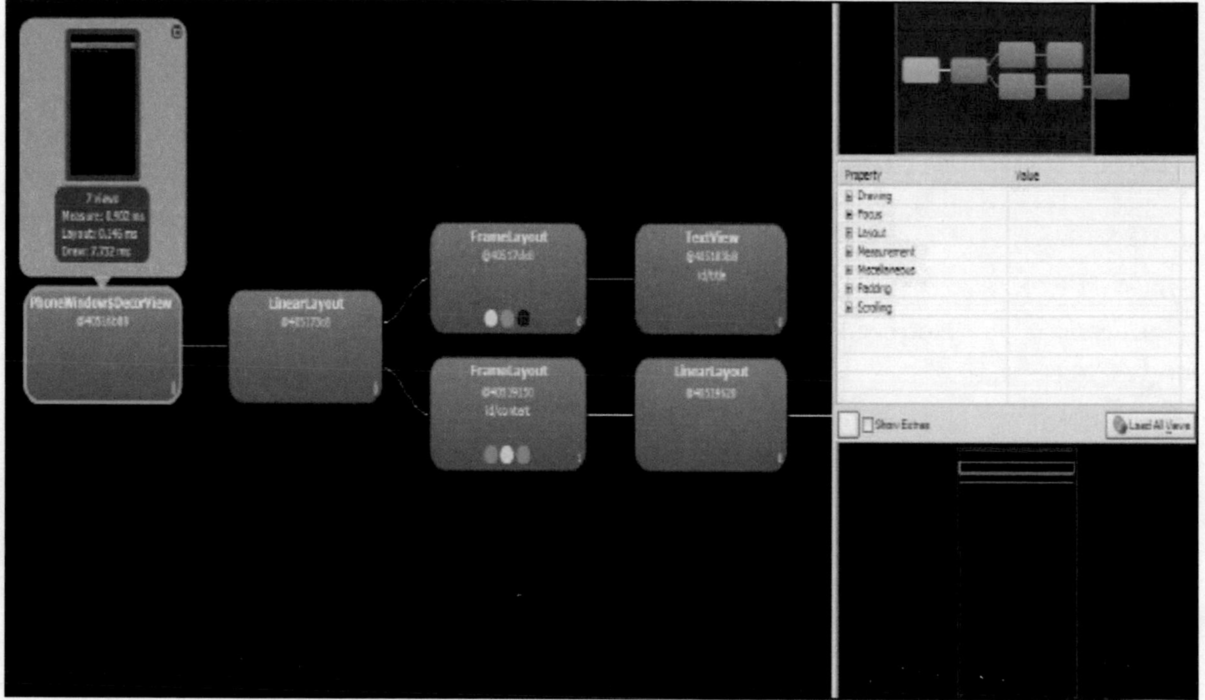

Figure 3.5: Hierarchy Viewer.

Table 3.1: Some Useful Android References.

Open Handset Alliance Website	`http://www.openhandsetalliance.com`
Android Developers Website	`http://developer.android.com`
Android SDK Download	`http://developer.android.com/sdk/`
Eclipse Integrated Development Environment	`http://www.eclipse.org/downloads/`
Android API Reference	`http://developer.android.com/reference/packages.html`

Table 3.2: References for Some Useful Android Components.

Function	Component	Reference
User Interface Views and Layouts	ViewGroup	http://developer.android.com/resources/tutorials/views/index.html
Data Storage and Sharing	ContentProvider	http://developer.android.com/guide/topics/providers/content-providers.html
Notifications and Alerts	NotificationManager	http://developer.android.com/reference/android/app/NotificationManager.html
Application Resources and Localization	Resource Manager	http://developer.android.com/guide/topics/resources/index.html
Location-based Services	LocationManager	http://developer.android.com/reference/android/location/LocationManager.html
Customizing User Input	InputMethodService	http://developer.android.com/reference/android/inputmethodservice/InputMethodService.html
Phone Calls and SMS	TelephonyManager	http://developer.android.com/reference/android/telephony/TelephonyManager.html
Power Management	PowerManager	http://developer.android.com/reference/android/os/PowerManager.html

CHAPTER 4

Windows Phone

Windows Phone is Microsoft's mobile operating system that competes with other modern mobile platforms such as iOS and Android.

4.1 EVOLUTION: FROM WINDOWS MOBILE TO WINDOWS PHONE

Microsoft has a long history in the mobile operating system market. In 2000, it introduced Pocket PC based on the Window CE kernel. Pocket PC was a PDA OS rather than a smart phone OS because it did not have telephony functions. In 2003, Microsoft launched Windows Mobile for smart phones, which was the first attempt from Microsoft to provide a smart phone OS. Microsoft also added phone functions to Pocket PC. Therefore, Windows Mobile as an OS had two different versions: Windows Mobile for Pocket PC and Windows Mobile for Smartphone. Later, Microsoft renamed them as Windows Mobile Professional Edition and Windows Mobile Standard Edition.

After iPhone appeared in 2007 and as Android gained smart phone market shares with a lightning speed, Microsoft realized that it must dramatically change its mobile OS in order to keep up with the competition. Windows Phone, a brand new mobile OS, was revealed in early 2010, marking a major departure from its predecessor, Windows Mobile, embracing the newest mobile device technologies and eying the newest feature sets of competing operating systems. The main differentiators of Windows Phone with respect to Windows Mobile can be summarized as follows.

- *Shifting the target market.* Windows Phone is primarily aimed at the consumer market rather than the enterprise market.

- *Exploiting emerging multi-touch technology*. The UI was overhauled and changed to be a finger-friendly UI supporting multi-touch. Touch devices with Windows Mobile had only stylus touch UI.

- *Recognizing the value/impact of the App ecosystem*. Windows Phone was carefully designed as a relatively "closed" platform. This is unlike Windows Mobile, where users had access to its device file system via file manager and could install applications freely by simply copying binary files to the device. Windows Phone file system is not directly accessible to end users or third party developers. For consumers, Windows Marketplace is the only way to install applications.

- *Requiring stricter hardware requirement to minimize fragmentation*. Windows Phone was introduced with a rich minimum set of hardware specifications that it requires of its OEM phone manufacturers. For Windows Phone 7, those specs included 800x480 resolution, 5-megapixel or greater camera, DirectX 9 compatible GPU, capacitive touch with 4 or more contact points, assisted GPS and other sensors like accelerometer, compass, proximity and light.

4.2 METRO UI

A design language, codenamed Metro, is used as a basis for the Windows Phone user interface. Metro UI echoes the kind of simplicity found in visual language signage (e.g., in airports and metro systems). It is also an essential part of Windows 8 and is positioned to guide UI development across all of Microsoft's consumer products.

Windows Phone features live tiles on the Start screen. A Tile can be anything from favorite contact to favorite application. Compared to static icons on other smart phone platforms, those tiles are "live" and can be easily updated by the application itself or via push notifications from a server. There are two types of tiles: *Application Tile* and *Secondary Tile*. An Application Tile is created when a user pins an application to the Start screen, whereas a Secondary Tile could be created programmatically by an application based on interaction from the user. An application can have only one Application Tile, but it could have multiple Secondary Tiles. For example a user may pin two favorite cities from a weather application for later quick access. They can be viewed as *deep links* into the application. Applications (blue) and Secondary (yellow) tiles are shown in Figure 4.1 below.

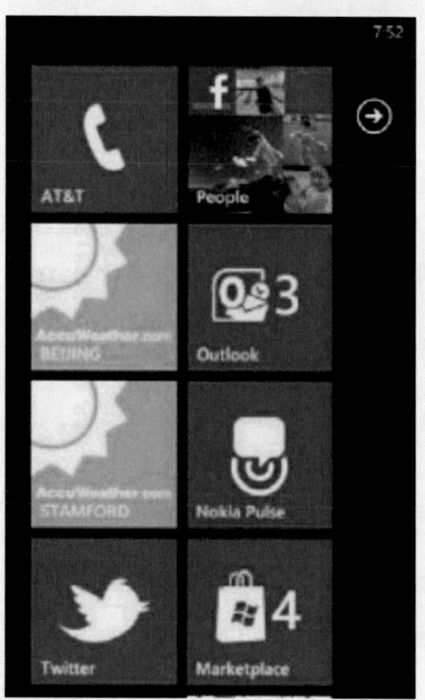

Figure 4.1: The Tile interface of Window Phone.

4.3 PLATFORM ARCHITECTURE

Windows Phone consists of four layers, as shown in Figure 4.2. Applications are running on top of the OS kernel, common base libraries and application frameworks. Silverlight and XNA are the two currently available frameworks for Windows Phone and will be discussed in Section 4.4.

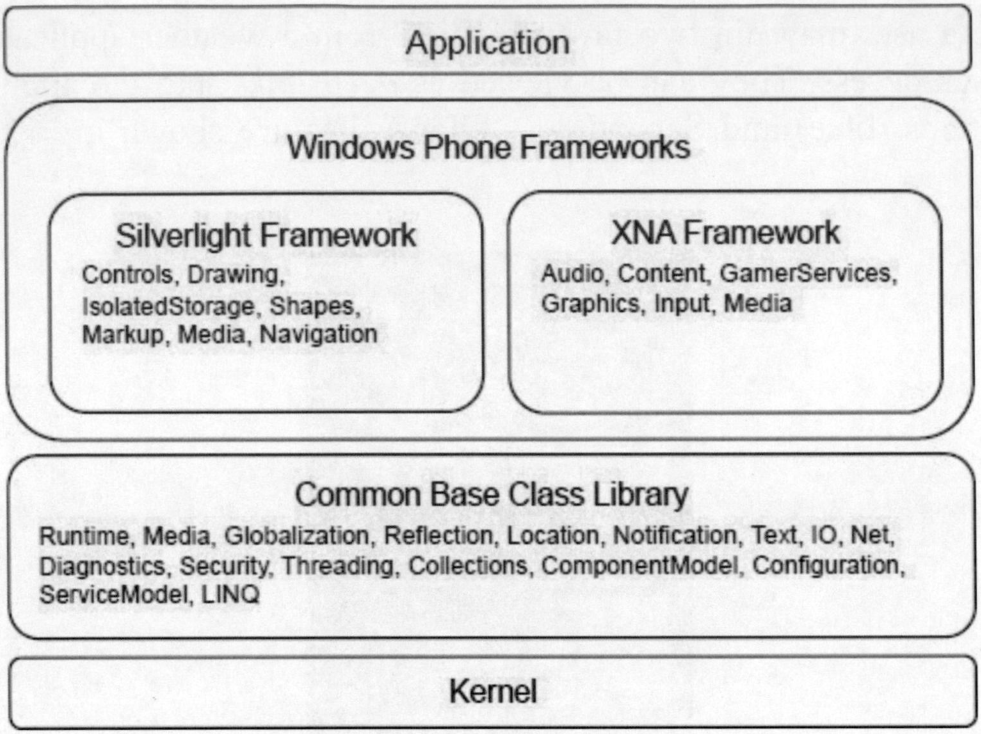

Application

Windows Phone Frameworks

Silverlight Framework
Controls, Drawing, IsolatedStorage, Shapes, Markup, Media, Navigation

XNA Framework
Audio, Content, GamerServices, Graphics, Input, Media

Common Base Class Library
Runtime, Media, Globalization, Reflection, Location, Notification, Text, IO, Net, Diagnostics, Security, Threading, Collections, ComponentModel, Configuration, ServiceModel, LINQ

Kernel

Figure 4.2: Windows Phone platform architecture.

4.3.1 THE KERNEL

Windows Phone 7 is based on the Windows Embedded CE kernel 6.0 R3, with additional functionality and features incorporated on top of the platform. All required drivers, file systems, network, graphics/rendering, and the phone update system run in kernel space. In the future, Windows Phone might run on the same kernel as the PC and tablet version of Windows.

4.3.2 COMMON LIBRARY

The common base class library includes support for runtime, globalization, multimedia, reflection, location, notification, text, I/O, networking, diagnostics, security, threading, configuration management, collections, component model, service model, and LINQ (Language Integrated Querying - a Microsoft .NET Framework component that adds native data querying capabilities to .NET languages). Below, we discuss a subset of the common base class library.

I/O

Windows Phone applications do not have direct access to the underlying kernel file system. All I/O operations are restricted to *isolated storage*, which is a data storage mechanism that provides isolation and safety by exclusively associating code with saved data. One application cannot access the isolated storage of another. A Windows Phone application can use isolated storage in the following ways:

- *Settings* (IsolatedStorageSettings): to store simple key/value pairs;
- *Files/Folders* (IsolatedStorageFile): to store data through file and folder creation;
- *Relational data* (LINQ to SQL): to store relational data in a local database.

Media

Windows Phone provides a unified programming model for building rich user experience that incorporates graphics, animation and media. Media library supports a variety of media codecs and provides access to the media available on the device as part of the music+video hub and corresponding Zune service. MediaElement control can hold audio or video clips and allows full control of a player's play, stop, pause and seek behavior, as well as stream buffering, download progress and volume control. Digital Rights Management (DRM) enabled media are enabled and managed under the Silverlight framework. More information about Media on Windows Phone can be found at `http://msdn.microsoft.com/en-us/library/ff402550(v=vs.92).aspx`.

Tasking

Similar to isolated storage concept, Windows Phone applications do not have direct access to the built-in applications on the device. *Launcher* and *Chooser* provide an indirect way for one application to call another.

Launcher is an API to launch a built-in functionality to accomplish some task. One simple example is PhoneCallTask, which can be used to make a phone call from an application. Chooser is an API to accomplish a task and return back the data that is chosen by the user to an application. For example, PhoneNumberChooserTask is used to choose a contact's phone number.

Push Notification Service

The Microsoft Push Notification Service in Windows Phone offers developers a robust channel to send information and updates to a mobile application from a web service. There are three kinds of notifications:

- *Toast* notifications display as overlay messages on top of the screen;
- *Tile* notifications drive a change to a tile on Start screen, with a title text, an image and a counter;
- *Raw* notifications send a small amount of data directly into the application. To receive raw notification, the application needs to be running at the time the notification is sent.

Location Service

Microsoft Location Service provides a single API to developers to get user's location information per user permission. This API hides the underlying details such as GPS, WiFi or cellular ID. Developers can set a desired accuracy and get user's current location, location change events, heading and speed among other parameters.

4.4 PROGRAMMING LANGUAGES AND FRAMEWORKS

Developers have two programming languages to choose from for writing Windows Phone applications: C# and Visual Basic .NET (VB.NET). C# is a modern object-oriented programming language introduced by Microsoft in 2000 for its .NET Framework. Its initial design was highly inspired by Java. However, in recent years, C# has been leading in many aspects such as LINQ and functional programming techniques. VB.NET can be considered as an object-oriented evolution of the classical Visual Basic language for the .NET Framework. Besides syntactical differences, C# and VB.NET are equally powerful for Windows Phone programming. The source code is compiled into Common Intermediate Language (CIL) bytecode whose execution is managed by the .NET Framework Common Language Runtime (CLR). Similar to the Java Virtual Machine (JVM), one of the benefits of CLR is that it provides garbage collection, which frees developers from worrying about memory management details (e.g., reclaiming memory occupied by objects that are no longer in use by the system).

In addition to programming language choices, Windows Phone offers two framework choices for application development, namely, Silverlight and XNA. The execution model for Silverlight is event-driven. Applications handle and process events based on user input or environmental effects. XNA uses a gaming loop and is focused on the content displayed on the screen. Typically, Silverlight is suitable for consumer or business applications while XNA is more suited for games development. From Windows Phone 7.5 onward, developers have a way to mix those two frameworks to deliver appealing user experience within a single application. Silverlight and XNA are described in more details in the next sections.

4.4.1 THE SILVERLIGHT FRAMEWORK

Application Logic and UI

The Silverlight framework is designed to allow developers separate application logic naturally from its UI presentation. Extensible Application Markup Language (XAML) is an XML-based declarative language used in Silverlight to create UI elements, such as controls, shapes and other contents presented on the screen. XAML defines the visual appearance of a UI, and an associated code file defines the logic. Developers can often adjust the UI design without necessarily making changes to the logic in the code.

In addition to Windows Phone applications, XAML can be used to develop desktop applications. Competing technologies are MXML for Flex from Adobe and FXML for JavaFX from Oracle.

A sample XAML code to create a button on a screen is shown below:

```
<Grid x:Name="LayoutRoot" Background="Transparent">
  <Button Name="ClickMeButton" Height="72"
    Width="160" Content="Click Me" />
</Grid>
```

XAML elements such as `<Button>` are the equivalent of object instantiation in procedural code. Therefore, once an object is declared in XAML, there will be an object (`ClickMeButton` in our example of `<Button>` element) that can be used in the application C# or VB.NET code.

UI components on the screen have parent/child relationship. For example, the `<Button>` XAML code creates a button within the LayoutRoot Grid. In addition to the visual relationship, UI components can also have an effect on how events are processed. For instance, developers can, if they wish, attach an event handler to LayoutRoot Grid to handle the mouse button event of `ClickMeButton`.

Application Life Cycle

Events and states together make up for an application's life cycle in Silverlight. A Windows Phone Silverlight application has four events and three states, as shown in Figure 4.3 below (events in rectangles and states in circles).

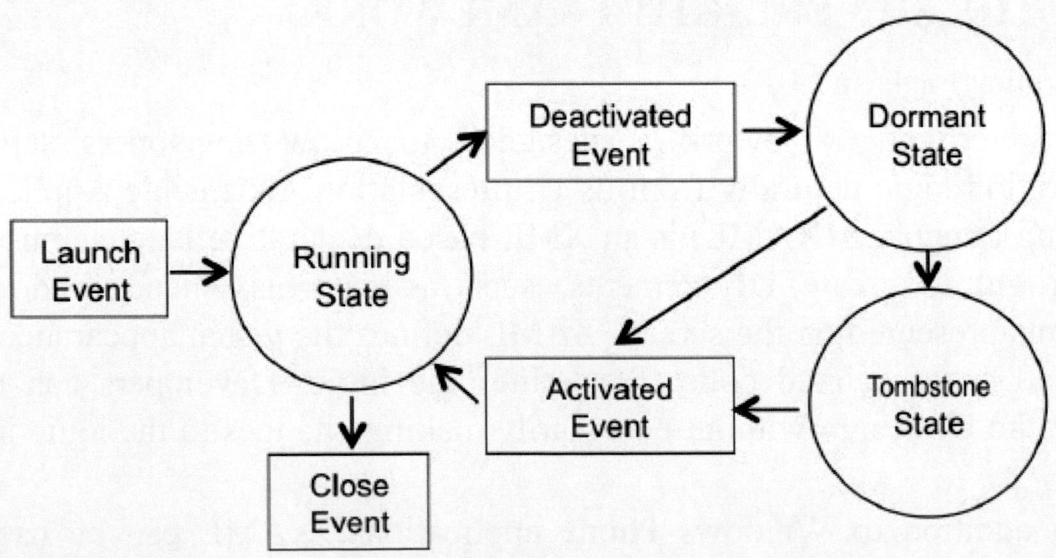

Figure 4.3: Silverlight application life cycle.

A Launch event is raised when the user starts an application. To provide best user experience on application launching, it is a good practice not to load heavy resources in this event.

```
private void Application_Launching(object sender,
                                    LaunchingEventArgs e)
```

Once an application is launched it goes to Running state. If the user navigates away from the application, a Deactivated event is raised.

```
private void Application_Deactivated(object sender,
                                    DeactivatedEventArgs e)
```

A Dormant state was added with the release of Windows Phone 7.5. In this state, application processing is stopped even though the application stays in memory to allow for fast switching (resumption). If the system finds another running application need more memory, it would tombstone a dormant application by pushing it to Tombstone state.

When the user switches back to a dormant or tombstoned application, Activated event is raised and code provided by the developer for handling application resumption is executed.

```
private void Application_Activated(object sender,
                                    ActivatedEventArgs e)
{
   if(e.IsApplicationInstancePreserved == true)
   {      // Dormant State
   }
   else
   {      // Tombstoned State
   }
}
```

Following the recommendation of Metro UI, applications should not have an explicit exit button, instead, a Close event is fired when the user is on the first screen of the application and navigates back. On this event, the system terminates the application after executing code provided by the developer to save important application (state) data within a short window of time.

```
private void Application_Closing(object sender,
                                  ClosingEventArgs e)
```

4.4.2 THE XNA FRAMEWORK

The XNA Framework is a managed runtime environment for game applications on top of the .NET Framework. With XNA, other different platforms can be targeted and accessed including Windows PC, Xbox and Windows Phone. The XNA platform saves game developers from writing boilerplate code and brings different aspects of game production into a single system.

Essentially, XNA provides a complete set of managed APIs for game development, including 2D sprite-based APIs that support rotation, scaling, stretching, and filtering as well as 3D graphics APIs for 3D geometry, textures, and standard lighting and shading.

There are four phases for any XNA game application, as described below and shown in Figure 4.4.

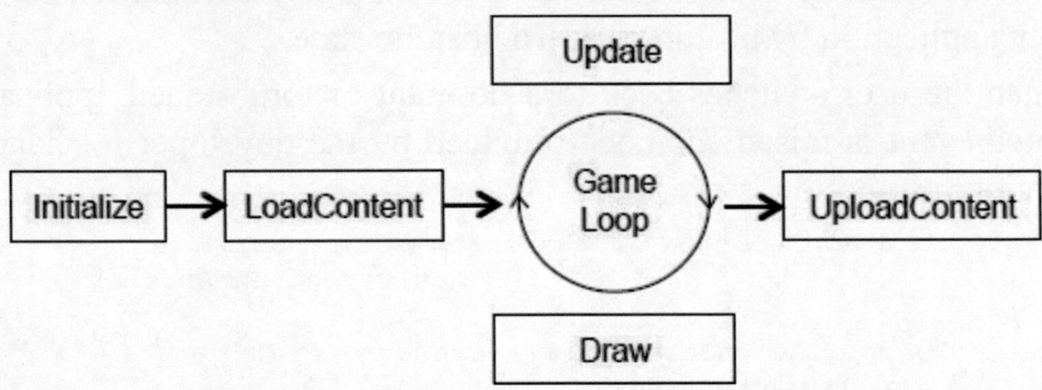

Figure 4.4: XNA game application phases.

First: Initialize()

This is where non-graphical initialization code for the game is placed.

Second: LoadContent()

After initialization, the game loads content such as sprites, images, sounds, and other resources of the game. With larger projects one may need to consider writing a more elegant solution to load the content only if and when it is needed.

```
protected override void LoadContent()
```

Third: Game loop with Update() and Draw()

After LoadContent(), an XNA game enters the game loop, which consists of two processes that run asynchronously:

- Update(): XNA automatically calls this method every update cycle. The game logic should reside here to manage tasks such as collision handling, inputs handling and game world updating.

```
protected override void Update(GameTime gameTime)
```

- Draw(): Draw the updates to provide visual feedback. XNA uses the full screen all the time, and the developer must generate full screen content for display within the code.

```
protected override void Draw(GameTime gameTime)
```

Fourth: UnloadContent()

The UnloadContent method is used to release all game assets prior to application exit.

```
protected override void UnloadContent()
```

4.4.3 MIXING XNA WITH SILVERLIGHT

Very often developers wish to combine XNA and Silverlight components into one single project. For instance, in an XNA game Silverlight could be used to provide game menu, or in a business application XNA could be used to provide 3D animation. With the release of Windows Phone 7.5 onward, Silverlight and XNA Frameworks can be combined into the same application by using the new SharedGraphicsDeviceManager and the UIElementRenderer class.

4.5 DEVELOPMENT TOOLS

Several development tools and IDE's are available for Windows Phone application developments. Each of those tools will be described briefly in the following sections.

4.5.1 VISUAL STUDIO

Visual Studio is Microsoft's IDE for developers to create Silverlight or XNA applications for Windows Phone. It can be used to create, develop, debug or package projects, and to generate package manifests. In addition to the paid version, Microsoft provides a free version called Visual Studio Express for Windows Phone. A performance analysis tool within Visual Studio enables developers to find out where the program is spending most its time and resources, which assists developers to analyze and optimize their code.

4.5.2 EXPRESSION BLEND

Expression Blend for Windows Phone is an integrated design environment which allows designers to create XAML-based interfaces for Windows Phone applications. It shares the same project structure with Visual Studio, which allows easy collaboration between designers and developers.

4.5.3 WINDOWS PHONE EMULATOR

Windows Phone Emulator is integrated into both Visual Studio and Expression Blend. It fully supports application deployment, debugging and execution. It supports GPU emulation, multiple touch points (pinch and zoom on multiple touch monitors), camera emulation and GPS and accelerometer sensors emulation. It also emulates several phone behaviors like placing a call and sending an SMS message. It does not, however, emulate actual hardware speed nor does it offer a real device experience. It provides built-in Internet Explorer browser application.

4.5.4 XNA GAME STUDIO

The XNA Game Studio is an integrated design environment that developers can use to build games for Windows, Xbox and Windows Phone. It extends the Visual Studio tools to support the XNA Framework, and includes tools for incorporating graphical and audio content into the game application.

4.5.5 SILVERLIGHT FOR WINDOWS PHONE TOOLKIT

Silverlight for Windows Phone Toolkit is a free, open source project offering developers additional controls for Windows Phone application development.

4.5.6 APP HUB AND MARKETPLACE

Microsoft App Hub offers tools, sample code, community support and educational resources for Windows Phone development. It provides a dashboard to developers so that they can manage all aspects of the application submission cycle, monitor downloads and track how much money is earned for paid applications.

Once an app is certified and approved on the App Hub, it can be published automatically or manually based on the developer's preferences in the Windows Phone Marketplace, which is the application store where Windows Phone users can browse, purchase and download applications to their Windows Phone devices. End users can either download applications directly from the Marketplace client on Windows Phone, or download from Marketplace on their PC and applications will be automatically pushed and installed on their devices.

From the App Hub, developers can also select Private Distribution Service, which allows developers to privately distribute certified apps to a targeted group of users. This distribution mode is a perfect enabler for deploying enterprise applications (that are not targeted to the general public) using the same infrastructure used for consumer apps.

4.5.7 WINDOWS AZURE TOOLKIT

The Windows Azure Platform is Microsoft's cloud platform used to build, host and scale web applications through Microsoft's data centers. To make it easier for Windows Phone developers to use Windows Azure, Microsoft created the Windows Azure Toolkit for Windows Phone, which provides a set of Visual Studio project templates that give the developer a starting point to build Windows Phone applications tied into services running in Windows Azure cloud. The toolkit also includes libraries, sample applications, and other help documentations.

Table 4.1: Some Useful Windows Phone References

Windows Phone Developer Documentation	`http://msdn.microsoft.com/en-us/library/ff402535(VS.92).aspx`
Windows Phone Class Library Reference	`http://msdn.microsoft.com/en-us/library/ff626516(v=vs.92).aspx`
Silverlight for Windows Phone Toolkit	`http://silverlight.codeplex.com`
Windows Azure	`http://www.windowsazure.com`
Windows Azure Toolkit for Windows Phone	`http://watwp.codeplex.com/`
Visual Studio 2010 Express for Windows Phone	`http://www.microsoft.com/visualstudio/enus/products/2010-editions/windows-phone-developer-tools`
Windows Phone Performance Analysis	`http://msdn.microsoft.com/en-us/library/hh202934(v=vs.92).aspx`
App Hub	`http://create.msdn.com`

CHAPTER 5

Mobile Web

5.1 MOBILE WEB EVOLUTION

The beginning of mobile browsers was marked by the advent of the Wireless Application Protocol (WAP) by the Open Mobile Alliance (OMA). WAP 1.0 was released in 1998. It defined a markup language, Wireless Markup Language (WML) and a scripting language, WMLScript. WML and WMLScript were specially designed for wireless phones with built-in adaptations and consideration for the limited bandwidth wireless network, display limitations and modest CPU speeds of the phones at that time. Browsers supporting WML can still be found today. WAP 2.0 was released in 2002 also by OMA. The markup language of WAP 2.0 is XHTML Mobile Profile, which can be considered as a subset of XHTML. Similarly, WAP 2.0 also introduced a subset of the Cascading Style Sheet (CSS) standard named Wireless CSS or WAP CSS. The World Wide Web Consortium (W3C) also defined mobile HTML/CSS standards similar to WAP 2.0. However, OMA's standards are more widely implemented today.

Mobile Web has gone through dramatic changes over the past few years. It has evolved from a simple content viewing platform to an application platform. Today's modern mobile browsers are not limited to mobile specific standards. Typical smart phone browsers are able to render full desktop sites with HTML, JavaScript and CSS. They also lead in HTML5 implementation.

Mobile Web applications could be represented in one of two forms.

1. *Mobile Website.* Users access websites though mobile browsers. Application logic either exists on the server side, or on the client side as JavaScript code.

2. *Offline Web Application*. Some mobile platforms support web applications in downloadable forms just like native applications. This kind of applications is also called *Mobile Widget*. An offline mobile application is rendered and executed within a "chromeless browser," which is a browser without the address bar and menus, which gives the user the perception of a standalone application.

From the surface, mobile web application development is not dramatically different from desktop web development: they both use standard web technologies such as HTML, CSS and JavaScript. The main principles remain the same, especially the *progressive enhancement* design approach, in which multiple layers of the contents and functionality can be accessed through any browser or Internet connection, providing those with better bandwidth, more advanced and enhanced features. Mobile Web development typically starts out with semantic and well-structured HTML markup that works even on browsers without CSS or JavaScript support. Then the look and feel (styling) can be progressively enhanced with CSS. The last step is to add rich, interactive features (behavior) using JavaScript. Each step can be further enhanced: basic HTML with HTML5, CSS with hardware-accelerated CSS animations and transitions, and JavaScript with advanced JavaScript device APIs.

For a long time, Oracle Java ME (previously Sun Java ME/J2ME) has been the cross platform runtime for mobile phones that developers could use to write code once in the hope that it runs anywhere. However, today's smart phones do not need cross-platform runtime support as the Mobile Web is capable of providing the same level of portability through a set of web and mobile web standards, all shared by most smart phones. These standards include:

- For Desktop: general web standards: HTML 3.2/4.0 and XHTML 1.0/1.1, and future standards: HTML5 and CSS 3.0;

- For Mobile: OMA WAP 1.x (WML, WMLScript), OMA WAP 2.0 (XHTML Mobile Profile, WAP CSS), and W3C (XHTML Basic, CSS Mobile Profile).

5.2 BEING MOBILE FRIENDLY

Developing mobile applications under Mobile Web requires a delicate balance between a reserved, limiting development attitude and fancy, rich attitude of the same. On the one hand, writing mobile web applications is about "limiting yourself." Developers and designers have to pay special attentions to (and stay within limits of) mobile device scaled down capabilities such as screen size, network speed, CPU power, browser engine, the nature of interruptions, among other limitations. On the other hand, there are additional mobile-specific functions provided by HTML, CSS and JavaScript, which should be exploited freely to provide for a much richer user experience. In the following, we discuss several features and mechanisms that aim at making the web mobile friendly.

5.2.1 DEVICE DETECTION

To deliver a web application optimized for mobile devices, the first step is to detect the mobile device type from the browser's request. The device type identity can be determined from the User-Agent field transmitted as an HTTP header from the browser. For instance, the User-Agent of iPhone with iOS 5.0 is

```
Mozilla/5.0 (iPhone; CPU iPhone OS 5_0 like Mac OS X)
AppleWebKit/534.46 (KHTML, like Gecko)
Version/5.1 Mobile/9A334 Safari/7534.48.3
```

Precise detection of a mobile device is not an easy task as the number of mobile devices continues to grow. In an effort to ease the task of device detection, a number of free and commercial libraries/services have been developed. WURFL is an open source project which offers a device database in XML format and API libraries in both Java and PHP.

5.2.2 VIEWPORT META TAG

The Viewport meta tag informs browsers that the site is optimized for mobile. It gives more information regarding how content should fit on the device's screen. A mobile website should set the initial scale to 1.0 and disallow user scaling.

```
<meta name='viewport' content='user-scalable=no,
width=device-width, initial-scale=1.0,
maximum-scale=1.0' />
```

Even for desktop websites Viewport meta tags can be used to carry preferred width and initial scale so the site can have a better initial display on smart phone browsers.

5.2.3 CSS MEDIA QUERIES

With CSS3 Media Queries, developers can add expressions to media type to check for certain conditions (e.g., screen width, height, orientation, resolution) and apply different style sheets. For example the following statement applies the style sheet “small-devices.css” to a device with screen width less or equals to 480 pixels.

```
<link rel="stylesheet" type="text/css"
 media="only screen and (max-device-width: 480px)"
 href="small-device.css" />
```

5.2.4 ORIENTATION DETECTION

On smaller screens, keeping content in a single-column vertical layout is preferred for portrait mode. Since users are able to switch between landscape and portrait modes, thanks to onboard accelerometer, the web application should be designed to support both modes. For instance, under iOS and Android, one can detect screen orientation and orientation changes by listening for the window.onorientationchange event and querying window.orientation for the angle, respectively. For smart phone browsers that do not support device orientation events directly, one can listen for the window.onsize event and distinguish portrait vs. landscape modes by checking window.screen.height and window.screen.width.

5.2.5 TOUCH AND GESTURE EVENTS

As iOS and Android added touch event APIs to their browsers, a W3C working group started work on a specification for touch events. There are three basic touch events:

- touchstart: a finger is placed on a Document Oject Model (DOM) element;
- touchmove: a finger is dragged along a DOM element;
- touchend: a finger is removed from a DOM element.

At a higher level than touch events, iOS provides *Gesture Events*, which give access to predefined gestures containing scaling and rotation information. Both gesture events and touch events are sent during a multi-touch sequence.

5.2.6 OPEN NATIVE APPLICATIONS

HTML links can be used to launch phone calls or SMS functions as shown in the following statements:

```
<a href="tel:01234567890">Call us</a>
<a href="sms:01234567890">Text us</a>
```

Link tags can also be used to open other native applications. For instance, an iOS application can bind itself to a custom URL scheme. This scheme can be used to launch itself from either a browser or from another application. For Android, Intent can be registered as part of an HTTP URL or be activated from a MIME type. Windows Phone has similar features. For instance, URL scheme "`zune://navigate/?phoneappID=`" is used to launch Windows Marketplace client on the device.

5.2.7 DEVICE APIS

The W3C Working Group on Device APIs and Policy aims to create specifications for device JavaScript APIs, including file, contacts, calendar, messaging, media capture and system information. The scope of the work includes not only smart phones, but also tablets, TVs and other connected devices. Some of those APIs already exist in today's smart phones.

5.2.8 BROWSER FRAGMENTATION

One of the key challenges for web development is to ensure support for all different browsers. While a handful of browsers can be found for the desktop (Apple Safari, Google Chrome, Microsoft IE, Mozilla Firefox and Opera), a broader range of browsers is available for mobile devices. This ranges from low-end browsers on feature phones from various manufacturers, to high-end smart phone browsers that are capable to render full desktop websites. Some browsers are proxy based; meaning content rendering and logic processing are done on the server side. Examples of proxy-based browsers are Nokia Browser for Nokia Series 40 phones, Opera Mini, and UC Browser. Delivering mobile sites that work well across different mobile browsers is very challenging despite all standardization efforts. This fragmentation effect also challenged Oracle Java ME in the past where developers had to debug on every different Java ME device and make tweaks to support different devices (write once run everywhere never really materialized for mobile).

5.2.9 DATA OPTIMIZATION

Data optimization remains important for Mobile Web as wireless data services continue to have relatively limited bandwidth and higher latency compared to its fixed Internet counterpart. It is a good practice to set up the mobile application web server to automatically remove indentations and other unneeded empty spaces from HTML. There are tools available to compress or minify JavaScript and CSS files. Minifying JavaScript also serves another purpose and achieves an often desired benefit, code obfuscation. It would not add absolute security, but it does make it much harder for someone to steal and modify the application code.

Browsers send separate requests for each external file such as images, CSS and JavaScript files. To reduce the number of HTTP requests, multiple CSS or JavaScript files are often combined together before deployed. Data URI can be used to embed data, such as Base64 encoded images, directly into HTML or CSS.

5.3 HTML5

HTML5 is currently under development as the next major revision for the HTML Standard, which will replace both XHTML 1.1 and HTML 4. It is still in draft status, however some of the features have been already implemented in smart phone browsers. HTML5 today somehow refers to all new features of the forthcoming HTML5, CSS3 and the additional JavaScript APIs, combined.

Major features from HTML5 include:

- Better semantic tags such as `<article>`, `<section>`, `<nav>`, `<menu>`, and `<footer>`.

- The `<video>` and `<audio>` tags. These two tags add native support for embedding video and audio content in web pages, which traditionally has to be done with plug-ins like Flash.

- New form `<input>` types such as "email", "date", "tel", "number". On supported OS's, the device will display different corresponding form of soft keyboard which is powerful to the programmer and convenient to end users.

- CSS3 aesthetic features like rounded corners, gradients, shadows, transition, transform and animations.

- Canvas API, which brings full-control of the colors, vectors and pixels on the script to JavaScript.

- DOM Storage, which can be considered as an extension of the cookie function. It is a way to persist key/value pairs in a secure manner. The sessionStorage object is limited to the current browser session. The localStorage object is used to store data that spans multiple windows and persists beyond the current session.

- Application Cache (or AppCache) allows developers to specify which files the browser should cache and make available to offline users.

- Geolocation API. Strictly speaking Geolocation API is not part of HTML5 but rather a separate W3C standard. It provides a standard w to query user's location with JavaScript which is agnostic to the underlying sources of location information.

HTML5 has more features such as Web Workers and WebSocket, but many of those features are not widely implemented on mobile devices yet. More information about HTML5 can be found at `http://www.html5rocks.com`.

5.4 WEBKIT

WebKit is an open source browser engine originally developed by Apple and licensed through the LGPL and BSD licenses. The engine, which received contributions from many top companies such as Apple, Google and Nokia, has become the most popular Web engine that powers many browsers in the market to date (2012). Symbian was the first mobile platform to make use of WebKit in 2005. The following is an incomplete list of mobile browsers that are developed using the WebKit:

- Safari on iOS;
- Android browser;
- New browser on Blackberry;
- Symbian browser;
- Amazon Silk Browser for Kindle.

WebKit adds many extensions to CSS. Those extensions start with -webkit-. For example, the following device-pixel-ratio Media Query can be used to target the style for high pixel density displays.

```
<link rel='stylesheet' href='highRes.css' media='only
screen and
 (-webkit-min-device-pixel-ratio: 2)' />
```

Some of those WebKit extensions are making their way to be accepted as CSS3 standard. In CSS3 they are typically implemented without the WebKit prefix, for instance -webkit-border-radius vs. border-radius.

5.5 WEB VS. NATIVE VS. HYBRID

5.5.1 WEB VS. NATIVE

Compared to native applications, web applications have the following advantages:

- They are cross-platform. Every smart phone platform has a modern browser which supports standard web technologies. Although web applications may not be 100% portable on all smart phone platforms, they can always be designed in a way that maximizes their code portability.

- They are easy to deploy. Instead of going through various application stores operated by different platform providers, web application can be deployed directly to a server which makes them available immediately to the end user.

However, there are several reasons to opt to creating a native application:

- First, performance consideration. If the application is an immersive game that uses extensive 3D hardware acceleration, native application is a much better option.

- Second, access to special device features and attributes. Although there are JavaScript APIs for accessing device features, those APIs are not universally available on all platforms.

- Third, monetization. A native application with installation package is the way to go if the application is to be sold via an application store.

5.5.2 HYBRID APPLICATIONS

It is possible to take advantage of both web and native platforms by choosing to write hybrid applications. Typically web views can be included into native applications. Certain parts of the application, especially content view UI, can be written with web technologies. Smart phone platforms often provide APIs to make calls between native and JavaScript code. For

Android the class to include a web view into a native Java application is android.webkit.WebView, whereas iOS uses UIWebView, and Windows Phone uses WebBrowser control.

Adobe PhoneGap, which is an open source framework to build hybrid applications, is a good solution for creating applications that will run on various devices with the same code. In addition to wrapping web applications into a native form, PhoneGap provides a rich set of JavaScript APIs to access native phone functionalities such as file, sensors and media. PhoneGap supports iOS, Android, Windows Phone and other platforms.

5.6 OFFLINE WEB APPLICATION

We discuss a few features and standard specifications that aim at enabling offline web application processing.

5.6.1 HTML5 APPCACHE

AppCache is a new feature of HTML5. It uses a manifest (a simple text file) to determine which resources in the web application should be cached by the browser. To make the cached application appear closer to a native application, an application icon can be added to the application grid in iOS and Android based on a meta tag from HTML. An example of offline HTML5 application is shown in Appendix A.4.

5.6.2 W3C WIDGET

The goal of W3C Widget Packaging and Configuration specification is to propose a standard method for building and packaging widgets. Widgets are just ZIP packages containing at least two files: Configuration file (config.xml) and resource files (HTML, CSS, JavaScript and image files). The configuration file contains information about the widget such as name, version, author, icon, etc.

5.6.3 WAC

The Wholesale Applications Community (WAC) is an open global alliance that is creating a unified and open platform based on standard Web applications. WAC's member includes the world's largest mobile operators and manufactures. WAC's promise is to allow developers to write applications that are usable across a variety of devices, OS's and networks.

The WAC platform builds on the work of the BONDI project, the Joint Innovation Lab (JIL) device APIs and the GSM Association's OneAPI program. It uses a subset of the W3C Widgets 1.0 Packaging and Configuration specification while providing proprietary extension through its own WAC namespace.

5.7 MOBILE WEB APPLICATION FRAMEWORKS

jQuery Mobile and Sencha Touch are two popular mobile Web frameworks that represent two different approaches for mobile web application development. jQuery Mobile is based on jQuery and jQuery UI foundation. It provides a set of easy-to-use APIs for DOM traversing and manipulation, event handling, animation, advanced effects and AJAX (the Asynchronous JavaScript and XML programming methods). Sencha Touch is the mobile counterpart to the Ext JS framework. It provides a rich set of UI components, storage and data binding facilities using JSON and HTML5 offline storage, and more. For jQuery Mobile, developers write regular HTML as presentation layer of the content, then add jQuery Mobile to provide enhancements such as effects and animations. However, for Sencha Touch developers create interface programmatically with JavaScript. Both are powerful frameworks that can save time in creating mobile web applications.

5.8 DEVELOPMENT TOOLS

Unlike native application development, Mobile Web application development does not have "official IDEs." Some developers prefer plain text editors such as Notepad++. Some prefer full featured IDEs Based on the Eclipse Platform, Aptana Studio is an open source IDE for editing, previewing and debugging web and mobile web applications. Dashcode on

Mac from Apple can be used to create Mobile Web applications for Safari on iOS.

Device emulators typically have the same built-in browser as the real devices. Those emulators can be used to test Mobile Web applications on a given target platform. However, nothing can replace debugging and running the application on real handsets.

Selenium, the popular web browser automation tool, enables test automation on mobile browsers for Android and iOS via its WebDriver on those platforms. Finally, a handy tool called *weinre* (which stands for WEb INspector REmote) makes remote debugging possible on mobile devices, such as viewing and modifying the DOM, checking console.log messages, among other debug features.

Table 5.1: Some Useful Mobile Web References.

WURFL	`http://wurfl.sourceforge.net`
DeviceAtlas	`http://deviceatlas.com`
W3C mobileOK Checker	`http://validator.w3.org/mobile/`
W3C Device APIs Working Group	`http://www.w3.org/2009/dap/`
HTML5 Specification	`http://dev.w3.org/html5/spec/Overview.html`
Geolocation API Specification	`http://dev.w3.org/geo/api/spec-source.html`
HTML5 Rocks	`http://www.html5rocks.com`
The HTML5 Test	`http://html5test.com`

Mobile HTML5	http://mobilehtml5.org
WebKit	http://www.webkit.org
W3C Widget	http://www.w3.org/TR/widgets/
The Wholesale Applications Community	http://www.wacapps.net
PhoneGap	http://phonegap.com
jQuery Mobile	http://jquerymobile.com
Sencha Touch	http://www.sencha.com/products/touch
AJAX	http://en.wikipedia.org/wiki/Ajax_(programming)
Aptana Studio	http://aptana.org
Selenium	http://seleniumhq.org
weinre	http://phonegap.github.com/weinre/

CHAPTER 6

Platform-in-Platform: Location-Based Services (LBS)[1]

6.1 HISTORICAL PERSPECTIVE

The main origin of Location-Based Services (LBS) was the E911 (Enhanced 911) mandate, which the U.S. government passed in 1996. The mandate was for mobile-network operators to locate emergency callers with prescribed accuracy, so that the operators could deliver a caller's location to Public Safety Answering Points. Cellular technology couldn't fulfill these accuracy demands back then, so operators started enormous efforts to introduce advanced positioning methods. To gain returns on the E911 investments, operators launched a series of commercial LBSs. In most cases, these consisted of finder services that, on request, delivered to users a list of nearby points of interest, such as restaurants or gas stations. However, most users weren't interested in this kind of LBS, so many operators quickly phased out their LBS offerings and stopped related development efforts. It was 2005 before the LBS wind started blowing again—this time in the right direction. Several significant developments and favorable conditions came together at that time to resurrect LBSs.

The emergence of GPS-capable mobile devices, the advent of the Web 2.0 paradigm, and the introduction of 3G broadband wireless services were among the enabling developments. In the meantime, small software and hardware companies realized a broad range of LBS capabilities for both mass and niche markets and laid down the foundation for a new generation of LBSs. A timeline of the most significant developments and landmark events in the short history of LBS is depicted in Figure 6.1.

6.2 EVOLUTION OF LBS

Early LBS's were reactive, requiring user initiation of service requests. They were also self-referencing and single-targeted, meaning concerned only with one mobile user location. They were mainly content-oriented, providing only information based on location. Early LBS's were "operator" centered and owned. The network operator controlled the service definitions either directly or through partnerships with major content providers. For instance, Verizon's early LBS services (mid 1990's) were provided for a very short time in partnership with Microsoft MSN.

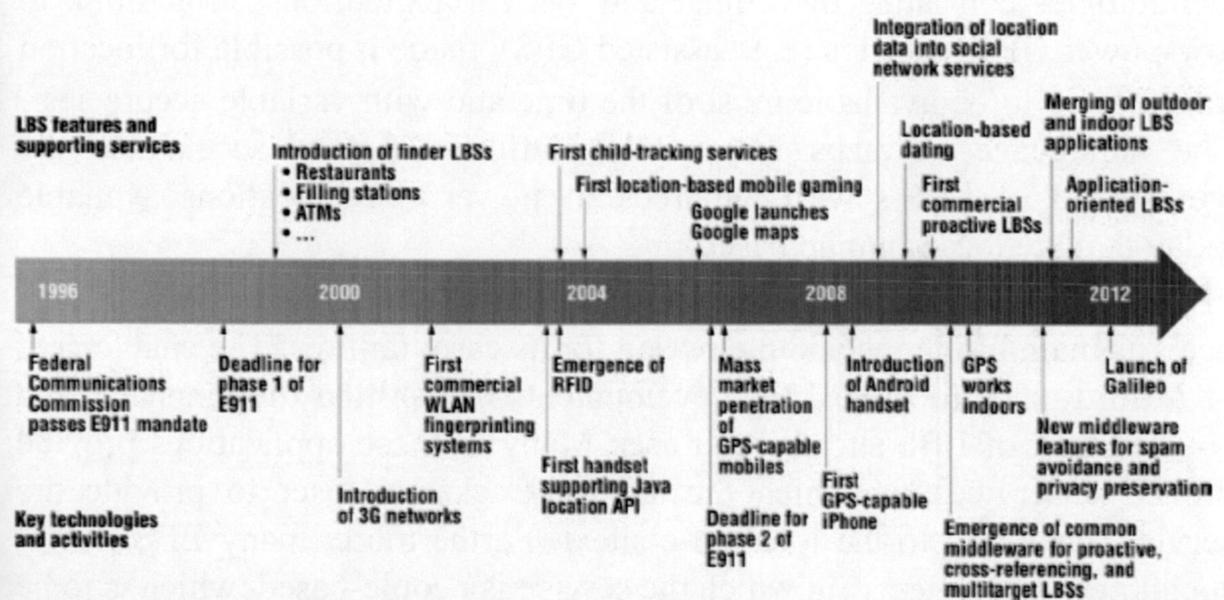

Figure 6.1: A brief history of Location-Based Services (Courtesy IEEE Pervasive Computing April-June 2008).

In 2004, operators and other providers started offering services for fleet management and for tracking children and pets—these were the first examples of *cross-referencing* LBSs. Initial versions of these services were based on cell-ID positioning using triangulation techniques, which suffered from low accuracy and were soon replaced by GPS. With the emergence of GPS-capable and programmable mobile phones (e.g., java 2 Micro Edition phones, and later on the iPhone), savvy users started to write small applications passing location data to a central server to make their location available to other users. Soon, these early initiatives turned into professional businesses that created a broad range of proactive and multi-target services—such as for mobile gaming, marketing, and health. These developments were accompanied by Web 2.0 in which location became another context item exchanged between the members of a social network, which was the origin for location sharing, a basic function of many of today's multi-target LBSs.

As the mobile platform programmability developed, GPS technology improved, and WiFi permeated the globe; conditions were ripe for third party location-based applications to proliferate. An overlay of geolocation technologies consisting of cellular and WiFi triangulations, in addition to low-power GPS receivers (e.g., assisted GPS), made it possible for location information to be available most of the time and with variable accuracies[2]. The emergence of maps for mobile platforms further accelerated the widespread of LBS's with hundreds of powerful applications available today that are location/map based.

In retrospect, the network operator centricity of LBS's seems to have been its main limitation and the reason for its early failures. The enablement of a third party to drive LBS developments has shifted the emphasis and "ownership" of LBS's to the end user. Many of these applications provide proactive services, in which the service tracks the user to provide the service according to the location context. Furthermore, many LBS's offer social network "effects" in which the service is people-based, which is to be contrasted with content-based. Figure 6.2 depicts the "big bang" of LBS's, which have exploded through proactivity, community orientation and user centricity.

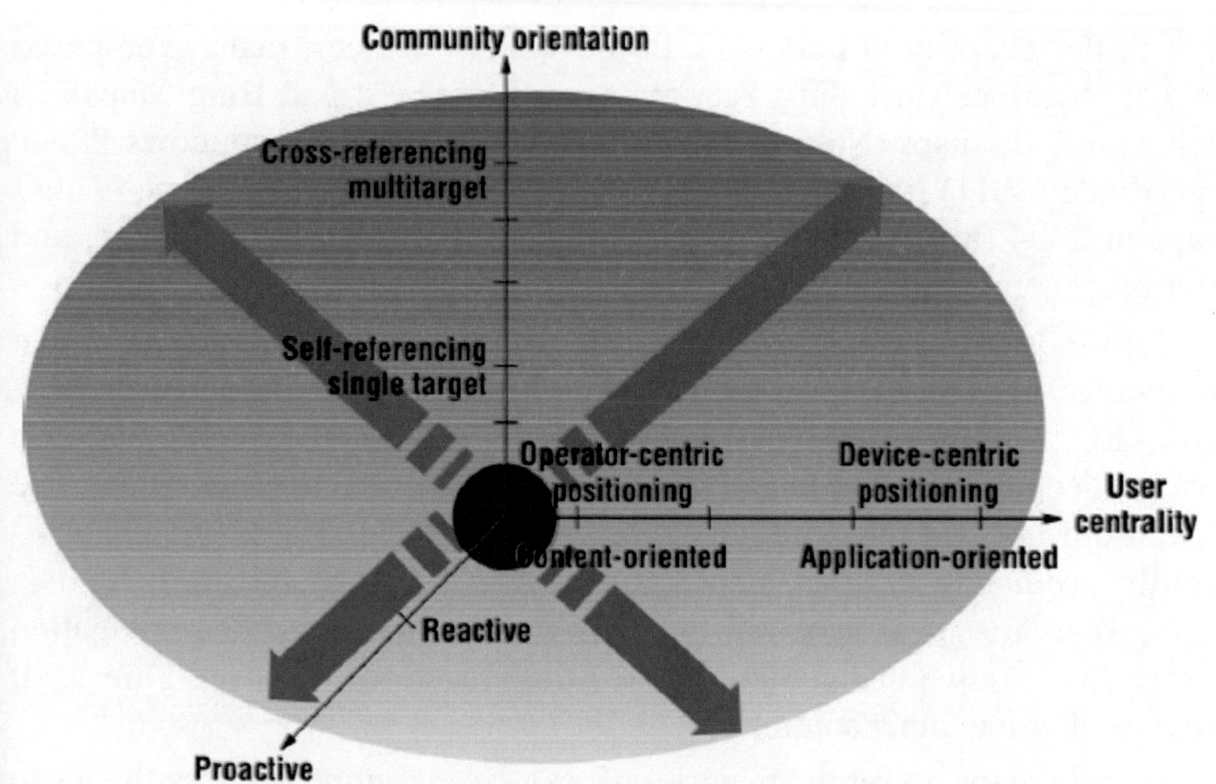

Figure 6.2: Location-Based Services evolution (Courtesy IEEE Pervasive Computing April-June 2008).

6.3 MAPPING THE WORLD

Interactive digital maps, in themselves, are a powerful and most natural user interface for LBS's. Expressing a target location is as easy as a few multi-touch interactions with the map. Finding the current location of a user cannot be expressed by any means better than the famous close-in circle on a map. For today's mobile platforms, "mobile maps" are an essential ingredient of the majority of LBS applications.

6.3.1 OUTDOOR MAPS

The world of digital navigable maps that we are getting accustomed to today (web or mobile) can be traced back to NAVTEQ, the most dominant company in geographic information systems and electronic maps. NAVTEQ was founded in 1983 under a different trademark name, and in 2007 was acquired and became a wholly owned subsidiary of Nokia. NAVTEQ powers the majority of portable GPS navigation devices, many web based map applications (including Yahoo! Maps, MapQuest and Bing Maps), as well as mobile maps (Nokia Maps, Bing Mobile Maps for Windows Phone and of late (2011) Maps for iOS). Even Google started off using NAVTEQ maps in 2004 (a service then called Google Local) before it switched, and later on generated its own map assets.

Tele-Atlas, a Dutch company established in 1984, was the main competitor to NAVTEQ. In 2008, and following the Nokia acquisition of NAVTEQ, Google—then a new competitor to Nokia in the mobile platform space—decided to abandon NAVTEQ and switched to Tele-Atlas. The switch did not last for more than a year, after which Google embarked on rapidly producing its own maps starting with the U.S. territory. Also in 2008, Tele-Atlas was acquired by Tom-Tom—a portable GPS navigation company—a move that confined Tele-Atlas' reach to only Tom-Tom itself along with a few other automakers.

Google maps power the majority of map-based online services thanks to its API, which allows a third party to embed Google maps on their web sites. The API is available for Mobile Web as a JavaScript as well as a collection of Web Services. An Adobe Flash API was briefly supported but has been dropped since 2011. In 2006, Google introduced Google Maps for Mobile which supported only J2ME at that time. Today, Google Maps for Mobile supports many platforms, including Android, iOS, Symbian, and Blackberry.

Many similarities in capabilities and functions (e.g., POI viewing, searching, route calculations, traffic, satellite images, etc.) can be found when comparing Nokia Maps with Google maps for mobile. However, Nokia Maps is pioneering in full offline mobile maps with turn-to-turn navigation. Its vector-based mobile maps application was initially released on Symbian OS. Now the support is expanded to other Nokia platforms and Windows Phone. Since the 5.0 release, Google Maps offers vector-based 3-D maps for Android with multi-touch gesture support. Dragging down with two fingers will tilt the map and switch between 2D and 3D views, while twisting fingers will rotate the map.

6.3.2 MAPPING THE INDOORS

As if the World Is Not Enough (1999 James Bond movie), the world's digital maps (web or mobile) are actively being augmented with geographical information about the indoors. Eying the magnificent retail, shopping and travel industries around the world, indoor maps are being created at a current pace estimated in the hundreds per month. This includes major airports, shopping malls, stadiums, resorts and other complex architectural spaces. This pace is bound to accelerate as indoor maps turn into a hot pursuit among competitors of this new pie of indoor LBS services.

In 2009, Micello introduced its first installment of indoor venue maps along with an iOS mobile navigation application. The maps were created manually utilizing publically available information in addition to specialized outsourced efforts whenever needed. The maps are overlaid nicely with Google maps, which makes it easy for users to switch between indoor and outdoor maps.

Micello maps allow for adding data layers to enable semantic descriptions, search and navigation of the indoor sub-spaces. For instance, a data layer in a mall could allow for "Men Shoes" and "restroom"search queries. An indoor Micello map of a mall is shown in Figure 6.3.

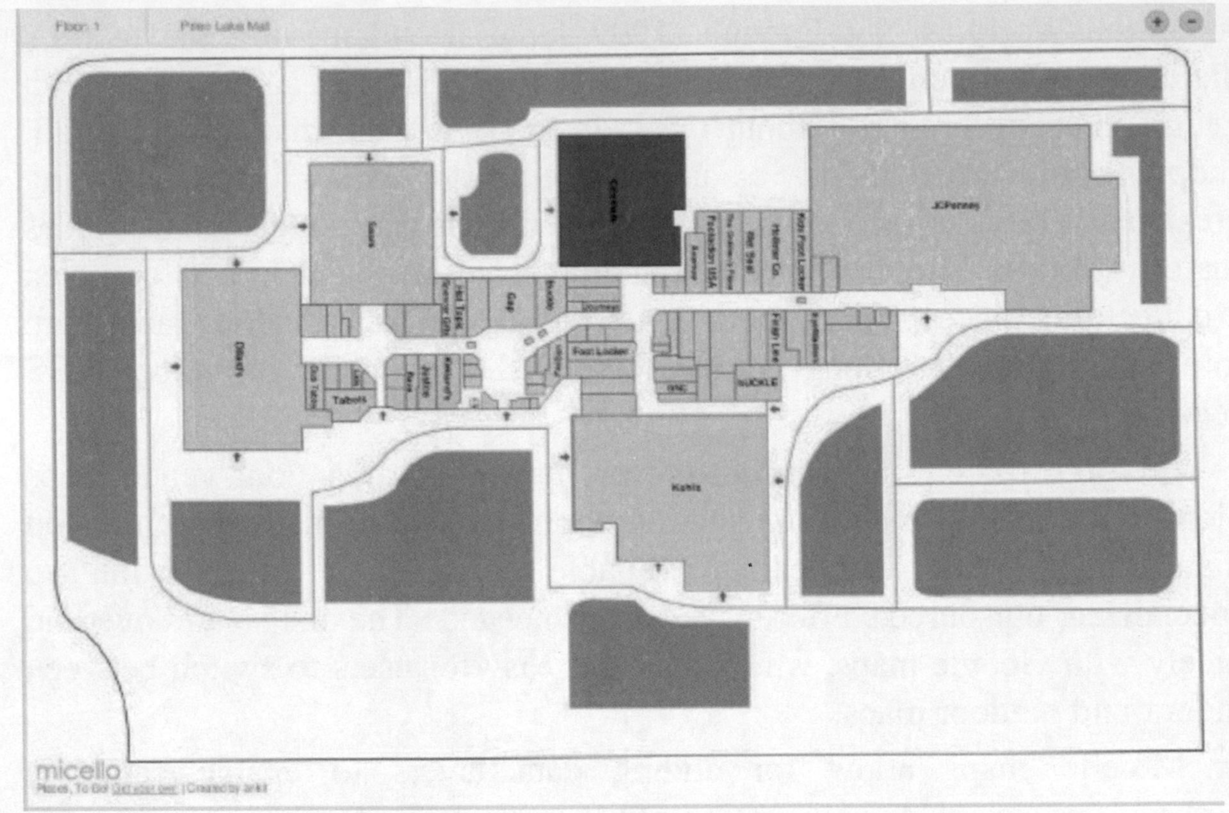

Figure 6.3: A Micello map of Prien Lake Mall, Louisiana.

Point Inside is another company offering indoor maps, but unlike Micello, its maps address the retail industry primarily with the goal of connecting retailers to their customers in several different ways. In addition to mobile applications (both iOS and Android), Point Inside offers a retailer platform—a set of tools to enable the retail store to deeply analyze their customers and provide highly personalized, timely services inside the retail store.

Indoor maps gained significant momentum in 2011 when Microsoft and Google officially launched their indoor maps and joined the race with game starters (or smaller players) Micello and Point Inside. Bing indoor maps were announced in August 2011 offering an initial installment of maps for some U.S. airports and shopping malls. The maps were made available to desktop users in August 2011, and have been available for mobile starting with Windows Phone 7.5. Bing indoor maps are seamlessly embedded and laid over Bing's outdoor maps, which requires no switching actions by the users—only zooming is required to see the details of an indoor map. A Bing indoor map on Windows Phone is shown in Figure 6.4.

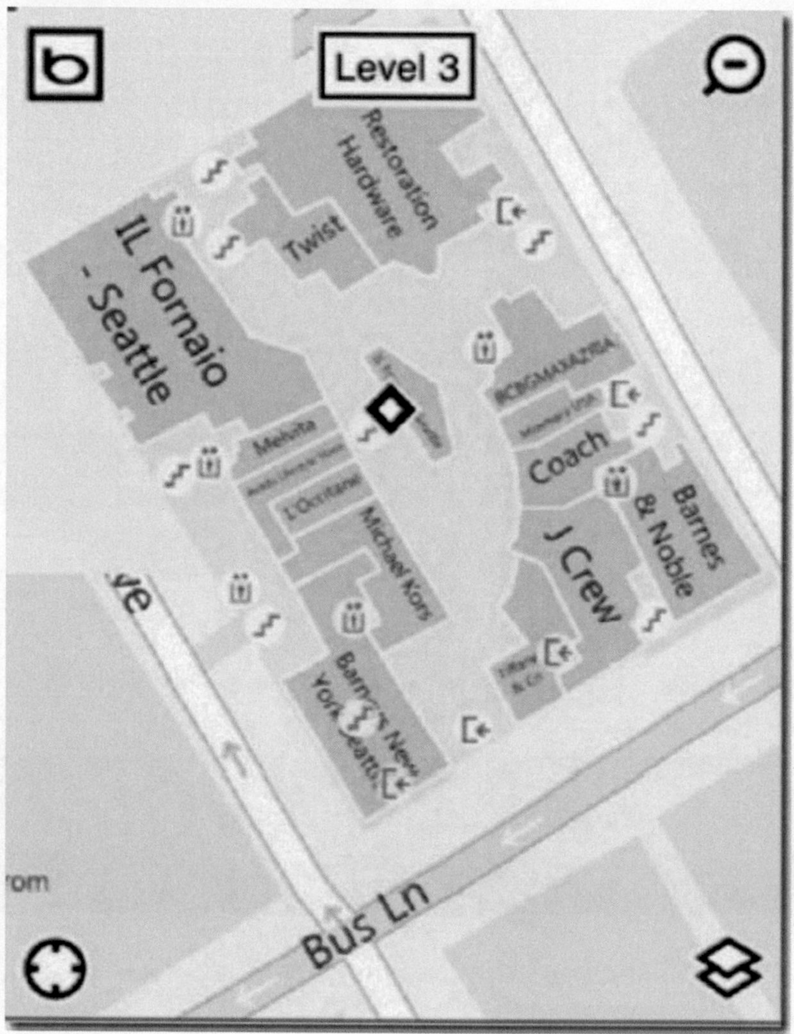

Figure 6.4: A Bing indoor map for Mobile.

Following Bing, the launch of Google Maps 6.0 for Android in November 2011 included indoor maps for several retail shops, malls and international airports in the U.S. and abroad. Google indoor maps are also seamlessly embedded. They also change automatically based on the floor the user is in. A comparison between two Google maps of the same location before and after the 6.0 release is shown in Figure 6.5

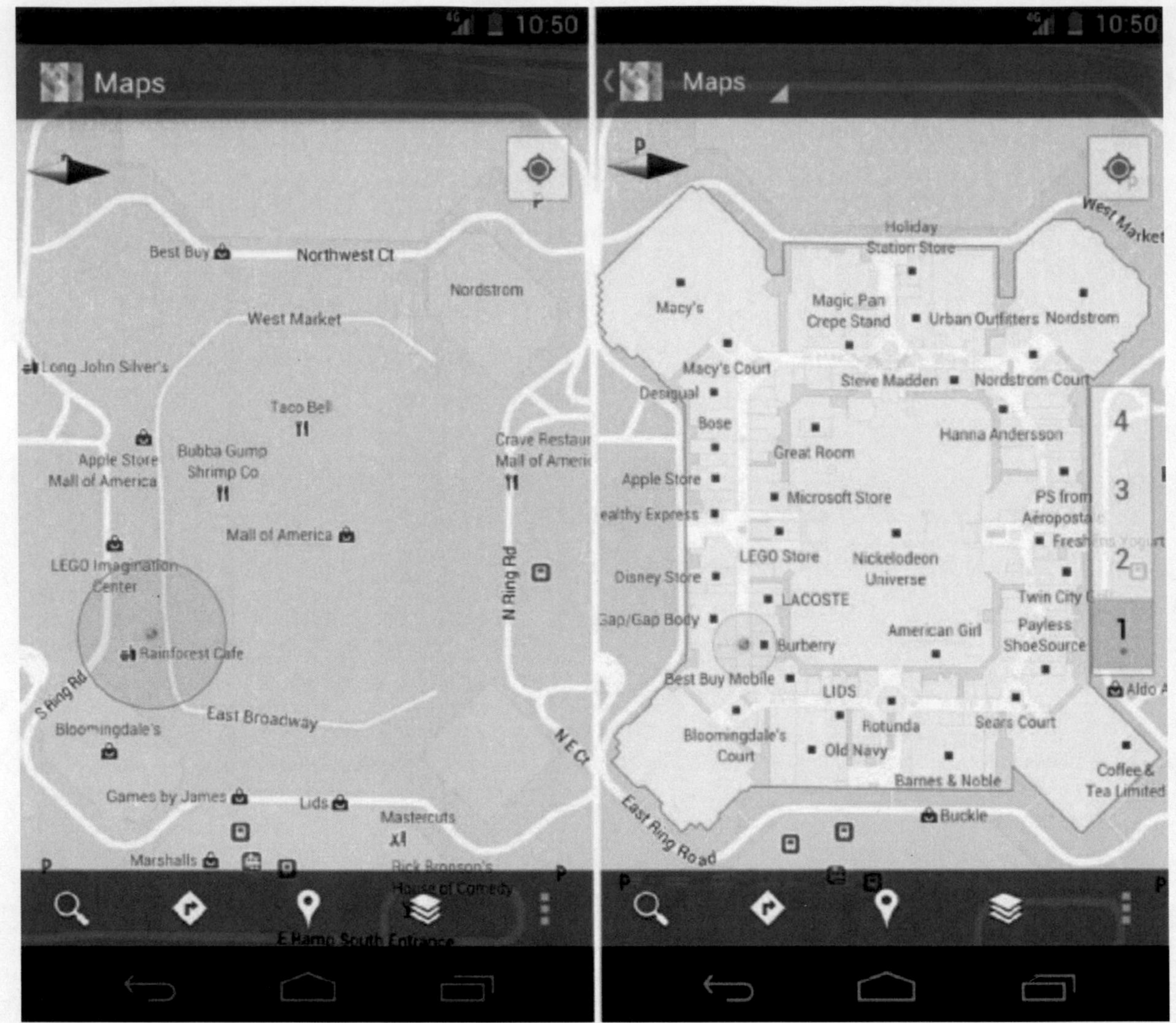

Figure 6.5: Mall of America in Minneapolis, before & after Google Map 6.0 for Android Release.

6.4 iOS LBS SUPPORT

Support for LBS on the iOS platform consists of a Core Location framework for accessing user location and getting notifications of location changes, a Map Kit framework for accessing and manipulating maps, and a "Maps" application that comes bundled in the platform for map viewing and browsing, and for searching and navigating to an address or a POI.

6.4.1 iOS CORE LOCATION FRAMEWORK

iOS makes use of WiFi and cellular networks in addition to GPS, and allows the developer to express the desired accuracy of the location information. The CoreLocation.framework must be linked to the application if location services are to be used. The framework empowers LBS applications with the following capabilities.

- **Verifying device capability for location services.** This allows LBS applications to make sure they can run successfully, as they start up. Also, App stores can use the same verification capability to ensure the app is downloaded to an appropriate device. This is done by including a UIRequiredDeviceCapabilities key in the application Info.plist file. Values for this key could be "location-services" for any available location service, or could be "GPS location-service" if high accuracy service is required. And even if the device is location service capable, and an LBS application successfully starts up, developers are provided with the LocationServiceEnabled class method to call prior to getting location information, to make sure location service is accessible. This is useful in cases such as when the user opts to disabling location services among other situations.

- **Obtaining current location.** The framework provides a configurable standard location service often used with most applications. To use this standard service, an instance of the CLLocation-Manager class should be created and configured with the desiredAccuracy and distanceFilter properties (required distance between two successive location updates). To begin receiving location notifications, a delegate must be assigned to the object and the startUpdatingLocation method must be called. As location data becomes available, the location manager notifies its assigned delegate object. The most recent location data can also be obtained directly from the CLLocationManager object without waiting for a new event to be delivered to the delegate. Code for starting a standard location service is shown below, courtesy the online iOS Developer Library.

```
- (void)startStandardUpdates
```

```
{
   // Create the location manager if this object does
   not
   // already have one.
   if (nil == locationManager)
      locationManager = [[CLLocationManager alloc]
      init];

   locationManager.delegate = self;
   locationManager.desiredAccuracy =
   kCLLocationAccuracyKilometer;

   // Set a movement threshold for new events.
   locationManager.distanceFilter = 500;

   [locationManager startUpdatingLocation];
}
```

Code for the delegate which receives the location updates is not shown, but should be provided and should implement locationManager:didUpdateToLocation:fromLocation: method which receives the update notification, and the locationManager:didFailWithError: method to receive notification if an error occurs.

- **Obtaining significant-change location**. A configurable, low-power, service for devices with cellular radios. This service provides location and location change events with accuracy acceptable by many applications. It does so at a much lower energy cost, which is an attractive feature to many "constant on" applications such as mobile social networks. The service does not use GPS, and instead utilizes the cellular network IDs. This service is available only on iOS 4.0 and later. To use the significant-change location service, an instance of the CLLOCAtionManager class should be created and assigned a delegate; then the startMonitoringSignificantLocation-Changes method should be called. As location data becomes available, the location manager notifies its assigned delegate object. As in the standard service, the most recent location data can be obtained directly from the CLLocationManager object without waiting for a new event to be delivered to the delegate. Code for starting a significant-change location service is shown below, courtesy the online iOS Developer Library.

```
- (void)startSignificantChangeUpdates
{
  // Create the location manager if this object does
  not
  // already have one.
  if (nil == locationManager)
      locationManager = [[CLLocationManager alloc]
      init];

  locationManager.delegate = self;
  [locationManager
  startMonitoringSignificantLocationChanges];
}
```

- **Region Monitoring.** In iOS 4.0 and later, applications can use region monitoring to be notified when the user crosses geographic boundaries. Classes are provided for defining region boundaries and for starting and receiving region boundaries crossing events (in or out).

6.4.2 iOS MAP KIT FRAMEWORK

Map Kit is an iOS 3.0 and later framework used to embed maps into applications through a fully functional map interface. The framework supports displaying street view and satellite view maps, search for addresses and POI, and zooming and panning through multi-touch display interactions. Currently, Google Maps are utilized by the Map Kit through a Google Map client. Hence, iOS developers using the Map Kit are bound by the Google terms of use and service associated with the Google Maps.

To use the features of the Map Kit framework, the MapKit.framework must be linked to the application in the Xcode project. The Map Kit provides interfaces for adding maps into an application area and for configuring the properties of the map. It provides a built-in graphical location finder right into the map (geocoding). It also provides a reverse geocoding capability. Detailed information about the various classes of the Map Kit framework is accessible from Apple iOS Developer Center.

6.4.3 OTHER LBS/MAPS SUPPORT FOR iOS

In addition to the Map Kit over Google Maps, Microsoft offers Bing Maps SDK for iOS. This SDK, released mid-2011, gives developers a set of Objective C classes to develop iPhone and iPad applications within the Xcode IDE. Included map control supports Bing's road, aerial and hybrid aerial map styles, and includes the ability to add pushpins to the maps and access the user's location via the GPS to locate the phone on the map. The Bing Maps SDK for iOS can be downloaded from Microsoft Download Center.

6.5 ANDROID LBS SUPPORT

Support for LBS on the Android platform consists of a Location Manager service, a Geocoding service, Google Map View—a Google MAP library for accessing and manipulating Google Maps through the Google API—and several powerful LBS applications that are included with the platform (e.g., Google Maps for Android, Places, Navigation, etc.).

6.5.1 ANDROID LOCATION MANAGER SERVICE

Android utilizes an overlay of location providers, currently including the GPS system, WiFi network and cellular network. This overlay provides alternative location sources to be used in different contexts (e.g., WiFi indoors and GPS outdoors). It also provides for a more reliable positioning when all sources are used combined, and gives the developer choices of positioning accuracy and power savings. Obtaining the user location reliably is the key service android provides for LBS development. To use the Android Location Manager Service, a Location Provider (GPS, WiFi, or Cell) and a location listener must be provided. Location is requested from the LocationManager by calling requestLocationUpdates(), passing it a LocationListener, which must implement several callback methods that the Location Manager calls when the user location or location provider status changes. The code segment below shows a call to the LocationManager with a simple listener (code courtesy of the Android Developer Guide).

```
// Acquire a reference to the system Location Manager
LocationManager locationManager = (LocationManager)
this.getSystemService(Context.LOCATION_SERVICE);

// Define a listener that responds to location updates
LocationListener locationListener = new
LocationListener() {
   public void onLocationChanged(Location location) {
    // Called when a new location is found by the
    network location provider.
    makeUseOfNewLocation(location);
   }

   public void onStatusChanged(String provider, int
   status, Bundle extras) {}
   public void onProviderEnabled(String provider) {}
   public void onProviderDisabled(String provider) {}
  };

// Register the listener with the Location Manager to
receive location updates
locationManager.requestLocationUpdates(LocationManager
.NETWORK_PROVIDER
, 0, 0, locationListener);
```

The first parameter in the call is the location provider type (NETWORK_PROVIDER, here, meaning WiFi or Cell), followed by the minimum time interval between notifications and the minimum change in distance between notifications—setting both to zero requests location notifications as frequently as possible. The final parameter is the LocationListener, which receives the callbacks for location updates.

Location updates can be obtained from more than one provider simultaneously, each using different frequencies. Best practice guidelines are recommended by the Android Developer Guide about the sequence of listening, caching and using of location information from the GPS system and the Network providers. Also user permission must be obtained when access is requested. An ACCESS_COARSE_LOCATION or

ACCESS_FINE_LOCATION permission must be provided in the application manifest or the application will fail to obtain the location information.

6.5.2 ANDROID GEOCODING SERVICE

Geocoding and reverse geocoding are another key services provided by Android to enable LBS applications. For Geocoding, an address is provided to obtain its longitude and latitude. The getFromLocationName Call (shown below) provides the location name and specifies the maxResults—the maximum number of geo references returned. A bounding box of latitude and longitude values can also be specified.

```
public List<Address> getFromLocationName (String
locationName, int
maxResults, double lowerLeftLatitude, double
lowerLeftLongitude, double
upperRightLatitude, double upperRightLongitude)
```

For reverse Geocoding, an array of addresses is returned for an area the surrounds a given latitude and longitude. The code for the getFromLocation is shown below.

```
public List<Address> getFromLocation (double latitude,
double longitude, int
maxResults)
```

6.5.3 GOOGLE MAP VIEW

The Google Maps library is used to create map activities in the application. It is not part of the standard Android library and must be included by updating the AndroidManifest.xml file. The first step to using a Google map view is to create a "MapView" layout for the application. This is done by creating or editing the layout XML file to include as the root node com.google.android.maps.MapView. Main.xml is shown below requiring an

API Key that must be obtained from Google prior to developing the map activity. Clickable indicates if the map is intended/permitted to be clickable by the user.

```
<?xml version="1.0" encoding="utf-8"?>

<com.google.android.maps.MapView

  xmlns:android="http://schemas.android.com/apk/res/
  android"

  android:id="@+id/mapview"

  android:layout_width="fill_parent"

  android:layout_height="fill_parent"

  android:clickable="true"

  android:apiKey="Your Maps API Key goes here"

/>
```

The second step is to construct the application by extending MapActivity. The latter is a sub-class of Activity provided by the Maps library, which includes all the capabilities needed to view and interact with the Google maps. Once constructed, the layout should be overridden by the layout of main.xml, which is the MapView layout. This is shown in the code segment below.

```
@Override

public void onCreate(Bundle savedInstanceState) {

  super.onCreate(savedInstanceState);

  setContentView(R.layout.main);

}
```

6.5.4 OTHER LBS/MAPS SUPPORT FOR ANDROID

Bing Maps Android SDK is an open source project that allows Android developers to use Bing Maps in their LBS applications. The SDK is in the form of a library that uses Bing Maps AJAX Control.

6.6 WINDOWS PHONE LBS SUPPORT

LBS support for Windows Phone consists of a Location Service API, Bing Maps and its Silverlight Control.

6.6.1 WINDOWS PHONE LOCATION SERVICE

To use the Location Service, WMAppManifest.xml—the manifest file of the application must include the location services capability ID_CAP_LOCATION in the <Capabilities> section.

The core of the location API is embodied in a class named GeoCoordinateWatcher in System.Device.Location namespace of the System.Device assembly. To retrieve user's location, the application needs to create an instance of GeoCoordinateWatcher, subscribe to events, and call it's Start() method (see the following code segment). The value passed to GeoCoordinateWatcher constructor specifies the desired accuracy. The two possible values are GeoPositionAccuracy.Default and GeoPositionAccurary.High.

GeoCoordinateWatcherprovides two events that the application can listen to:

- *GeoPositionStatusChanged* - indicates that the status of the Location Service has changed

- *GeoPositionChanged* - indicates that the latitude or longitude of the location data has changed

By listening to *GeoPositionStatusChanged*, the application can be notified of changes such as service disabled, enabled, initializing, etc., and communicate the current state of the service to the user (see the following code segment).

```
public MainPage()
{
  InitializeComponent();

  if (geoCoordinateWatcher==null)
  {
  // Create GeoCoordinateWatcher with high accuracy
  geoCoordinateWatcher = new
GeoCoordinateWatcher(GeoPositionAccuracy.High);
  geoCoordinateWatcher.MovementThreshold = 40;

  // Subscribe to status changed event
  geoCoordinateWatcher.StatusChanged +=
       new  EventHandler<GeoPositionStatusChangedEvent
       Args>
(OnStatusChanged);

  // Subscribe to position changed event to receive
  GPS coordinates
  geoCoordinateWatcher.PositionChanged +=
        New EventHandler<GeoPositionChangedEventArgs
        <GeoCoordinate>>
(OnPositionChanged);

  geoCoordinateWatcher.Start();
  }
}
```

```
private void OnStatusChanged(object sender,
GeoPositionStatusChangedEventArgs
e)
{
  switch (e.Status)
  {
    case GeoPositionStatus.Disabled:
      // The location service is disabled or
      unsupported
      break;
    case GeoPositionStatus.Initializing:
      // The location service is initializing
```

```
    case GeoPositionStatus.NoData:
      // The location service cannot get location data
      break;
    case GeoPositionStatus.Ready:
      // The location service is working and is
      receiving location data
      break;
  }
}
```

The PositionChanged event will not fire until the position of the device has changed as specified by MovementThreshold. There is no property or method that developers can set to determine how the location service is getting location data. But if data is from GPS, the HorizontalAccuracy and VerticalAccuracy properties will typically be a few meters; vs. non-GPS source, in which case, it could be 100 meters or higher. The following code segment shows how to obtain and extract location information in response to a position change event.

```
private void OnPositionChanged(object sender,
GeoPositionChangedEventArgs<GeoCoordinate> e)
{
  var location = e.Position.Location;

  // Access position information
  location.Latitude.ToString("0.000");
  location.Longitude.ToString("0.000");
  location.Altitude.ToString();
  location.HorizontalAccuracy.ToString();
  location.VerticalAccuracy.ToString();
  location.Course.ToString();
  location.Speed.ToString();
  e.Position.Timestamp.LocalDateTime.ToString();
}
```

6.6.2 BING MAPS CONTROL

To use Bing Maps control, the application needs a Bing Maps key, which can be created on the Bing Maps portal (https://www.bingmapsportal.com/) with a Windows Live ID. This line in the layout XML file creates a Bing Maps object:

```
<m:Map Name="bingMap" CredentialsProvider="YourBing-
Maps-Key"
Height="500" Width="450" />
```

Alternatively, a Bing Maps key can be added in the code.

```
bingMap.CredentialsProvider =
new ApplicationIdCredentialsProvider("Your-Bing-Maps-
Key");
```

In the XAML file of the page, the following namespace is required to add the Bing Maps control to the application.

```
xmlns:m="clr-
namespace:Microsoft.Phone.Controls.Maps;assembly=
Microsoft.Phone.Controls.
```

6.6.3 BING MAPS WEB SERVICES

Bing Maps GeoCode Service provides geocoding and reverse geocoding services that Windows Phone application can use. Bing Maps Route Service provides API to calculate route directions based on multiple locations on the map. The Bing Maps Interactive SDK provides sample code for interacting with those Bing Maps Web Services.

6.6.4 OTHER LBS/MAPS SUPPORT FOR WINDOWS PHONE

There is no official Google Maps client or Google Maps control for Windows Phone. But Bing Maps Control is so flexible that developers can actually bind Google Maps to it.

Nokia provides Nokia Maps and Nokia Drive clients that are exclusive to the Nokia brand of Windows Phone devices. Nokia Drive provides full offline, free turn-by-turn voice navigation which is a major differentiator from other Windows Phones.

6.7 MOBILE WEB LBS SUPPORT

The main avenue of support for LBS in Mobile Web applications is the emerging HTML5 standard, which is covered in Chapter 5. The main mechanism within HTML5 is a JavaScript API for map rendering and map control. Nokia Maps, for instance, offers a JavaScript API that supports HTML5, or rather takes advantage of HTML5 features to offer the API. Details of the API with numerous illustrative examples can be viewed from Nokia's Map API Online Guide. The Nokia Maps JavaScript takes advantage of:

- HTML5 <canvas> elements with or without hardware acceleration. Canvas, whose idea was introduced by Apple in 2004, offers a faster rendering of shapes and images than bitmaps. The Nokia Maps API detects canvas availability and attempts to render as much as possible of the map and its graphical annotations using canvas. The end result is an impressive rendering speed that comes close to native, non-web rendering.

- HTML5 allows for embedding of Scalable Vector Graphics (SVG) in HTML pages, which offers an alternative means to perform speedier rendering of shapes and images than bitmaps. The Nokia Maps API also supports the use of SVG especially for icons and custom markers. SVG follows the Document Object Model meaning the graphics consists of identifiable objects (e.g., a rectangle shape). Even though both are fast rendering mechanisms, the difference between SVG and canvas is in the lifecycle of the graphics: in canvas, the shape has to be redrawn (via JavaScript) along with the entire screen in response to any change. In SVG, only the shape object needs to be modified (also via JavaScript) to reflect any change of the object. For other changes external to the object, no changes need to be made.

- The W3C Geolocation API supported by HTML5. The API allows for the discovery of browser location. It makes no assumptions about the actual underlying location services that could be used to find the location. GPS, WiFi, cellular and even browser device IP address geo-referencing can be used individually or in any combination to implement W3C Geolocation API support in HTML5. While the original intention of the API is to use it for location tracking by web services of their clients (for service personalization and localization purposes), browsers can use the same API for positioning—finding their current location. The combination of access to Maps API and to the Geolocation API within HTML5 made it possible to create Mobile Web LBS.

Nokia also offers a powerful HTML5 Maps application for the Mobile Web. It supports pinch to zoom, routing for driving or walking, along with saving of favorite locations and POI information. The application supports offline mode, where users are able to download maps before their trips, which ensures availability of service regardless of Internet coverage. The application supports iOS 4.3+ and Android 2.2+ browsers.

In addition to outdoor maps, Mobile Web also enjoys access to indoor maps. For instance, Micello's JavaScript API for its indoor maps supports HTML5. Canvas is used extensively by the Micello API which is not surprising given the geometric shapes of indoor architectures (e.g., malls and airport terminals).

Table 6.1: Some Useful LBS References.

Micello	`http://www.micello.com`
Point Inside	`http://www.pointinside.com`
iOS Core Location Framework	`https://developer.apple.com/library/ios/#documentation/CoreLocation/Reference/CoreLocation_Framework/_index.html`
iOS Map Kit Framework	`https://developer.apple.com/library/ios/#documentation/MapKit/Reference/MapKit_Framework_Reference/_index.html`

Bing Maps for iOS	http://www.microsoft.com/download/en/details.aspx?id=1112
Android Location Manager	http://developer.android.com/reference/android/location/LocationManager.html
Android Google Map View	http://developer.android.com/resources/tutorials/views/hello-mapview.html
Bing Maps Android SDK	http://bingmapsandroidsdk.codeplex.com
Windows Phone Location Service	http://msdn.microsoft.com/en-us/library/system.device.location(v=vs.92).aspx
Bing Maps Portal	https://www.bingmapsportal.com
Bing Maps Geocode Service	http://msdn.microsoft.com/en-us/library/cc966793.aspx
Bing Maps Route Service	http://msdn.microsoft.com/en-us/library/cc966826.aspx
Bing Maps Silverlight Control Interactive SDK	http://www.microsoft.com/maps/isdk/silverlight/
Nokia Maps API	http://api.maps.nokia.com
W3C Geolocation API Specification	http://dev.w3.org/geo/api/spec-source.html

[1] Sections 6.1 and 6.2 of this chapter are based on an article published in the IEEE Pervasive Computing magazine (April-June issue of 2008) by Paolo Bellavista, Axel Kupper, and Sumi Helal.

2 A good source of information on location systems technologies is the Morgan & Claypool Synthesis Lecture entitled: "*Location Systems: An Introduction to the Technology Behind Location Awareness*", by Anthony LaMarca and Eyal de Lara. `http://www.morganclaypool.com/doi/abs/10.2200/S00115ED1V01Y200804MPC004`.

CHAPTER 7

The Future of Mobile Platforms

In this lecture, we covered the three major native mobile platforms of today—iOS, Android and Windows Phone—along with the device-agnostic HTML5 mobile web platform. We also covered Location-Based Services which can be considered as a platform in its own right despite requiring a device platform as host. Yet if one looks back, none of these platforms even existed commercially five years ago and mobile platforms such as Symbian and BlackBerry which ruled the roost in those days are declining at a fast pace. Interestingly, despite its steep decline, Symbian still continues to dominate the smartphone space in terms of the number of active users as of early 2012, but Android is currently the platform with the fastest growing market share and new subscribers. iOS has a strong base of followers and loyalists mainly due to its smooth user experience and unparalleled ease of use, while Windows Phone is the current underdog and relatively newcomer which is ranked somewhere between iOS and Android in terms of ease of use and customization.

Opinions are always highly polarized regarding which mobile platform will dominate the consumer space. However, regardless of which mobile platform rises to the top in the future, one thing can be stated for certain: the only thing constant in the mobile industry is change. In fact, one key observation that can be made is that all the mobile platforms started off as smartphones but proliferated into other device categories such as tablets and set-top boxes. As we look towards the future, we see a strong continuing trend towards convergence where the same platform will run across a broad range of device categories such as desktop, laptop, phone, tablet, gaming consoles and entertainment systems, including televisions. Even today, we have Apple iOS running on phones, tablets and set-top boxes while the Mac OS X for the desktop borrows a lot of user interface cues and other features (e.g., the App Store) from it. Android is more focused on phones and tablets with a more recent foray into set-top boxes through Google TV. However, it is Windows Phone, which despite being the youngest platform in the market, has been the most aggressive in its pursuit of convergence. Future versions of Windows are being designed from the ground up to run across multiple processor architectures with a touch-friendly and gesture-driven user interface, which seamlessly spans across different user experiences and device categories from the smartphone and tablet to the Xbox gaming console and the desktop PC.

Another trend that can be predicted as we look into the future is an imminent change of “screen status” for mobile platforms. Considered as “third screen” after Televisions and Computers, mobile platforms are now climbing to “first-screen” status in term of traffic and number of mobile devices. In response, many companies are reflecting this new reality by changing their business policies towards mobile from must-have and must-support, to target-first.

All in all, the future of mobile platforms looks brighter than ever before and the mobile industry is advancing innovation at a pace hitherto unseen.

APPENDIX A

TwitterSearch Sample Application

TwitterSearch is a simple application we developed for demonstration purposes in this lecture. It allows end users to search tweets from twitter.com via Twitter's REST Search API. To highlight the implementation differences, the same application was ported to each of the four mobile platforms we covered in this lecture namely, iOS, Android, Windows Phone and Mobile Web.

The TwitterSearch application user interface consists of two screens:

- Search Screen: Contains one text input box which allows the user to input keywords to search for on Twitter, and a search button.

- Results Screen: Presents the search results in a list, which shows the text of tweets containing the search item, the name of the tweet's author and the author's profile icon.

The URL for Twitter's Search API is as follows:

```
http://search.twitter.com/search.json?q=<string_of_searchterms>
```

The search result is returned in JSON (JavaScript Object Notation) format. JSON is a text-based, language-independent data interchange format. It is widely used in Web programming since it is lightweight as compared to XML and can be considered as a subset of the object literal notation of JavaScript.

To use this sample application, just input any keyword in the Search Screen text box and click on the "Search" button. The application will send the search request to the Twitter server using the above-mentioned URL, parse the JSON response and present the search results on the Results Screen.

We will now discuss the details of each platform implementation and present the associated application source code, which should be accessible in a zip archive format directly from the online version of this lecture. URLs for the source archive zip files are also listed at the end of this Appendix.

A.1 iOS CODE

With the introduction of iOS5, Apple has made significant changes in Xcode to streamline how an app developer can design and build his/her applications. A major area of focus has been to empower developers to create a large portion of their applications without writing a single line of code. iOS5 introduces the concept of Storyboards where a developer creates a visual representation of the UI flow between the different screens of the application (also known as, views). This is different from the Interface Builder tool used in previous versions of Xcode, which only allowed the visualization of individual views but required developers to write code to describe the transitions between them.

An iOS application is written in Objective C and described in an Xcode project file. Objective C class files have the extension '.m,' whereas the header files still retain the traditional C/C++ '.h' extension.

The main.m file is responsible for loading the application delegate class. The application delegate handles application-level events and allows message passing between different objects in the application in an easy manner without requiring sub-classing. It is always important to proactively ensure that the name of the application delegate class which is passed to the UIApplicationMain function is correct (in this case, TwitterSearchAppDelegate).

main.m

```
#import <UIKit/UIKit.h>

#import "TwitterSearchAppDelegate.h"

int main(int argc, char *argv[])
{
```

```
    @autoreleasepool
    {
      return UIApplicationMain(argc, argv, nil,
               NSStringFromClass([TwitterSearchApp
               Delegate class]));
    }
}
```

iOS applications follow the Model-View-Controller architecture where Model objects represent data, View objects handle the user interface and presentation of the data and Controller objects handle the interaction between the Views and the Models. An iOS application consists of one or more views where each view describes an area on the screen which is visible to the user. Each view also has an associated view controller of type UIViewController which handles interaction with the objects embedded in the view. The TwitterSearch application consists of two views: The Search View and the Search Results View (Figure A.1). The Search View is handled by the TwitterSearch View Controller (class name: TwitterSearchViewController) whereas the Search Results View is handled by the TwitterSearch Results View Controller (class name: TwitterSearchViewResultsController). The TwitterSearchViewResultsController class is of type UITableViewController which is a sub-class of UIViewController.

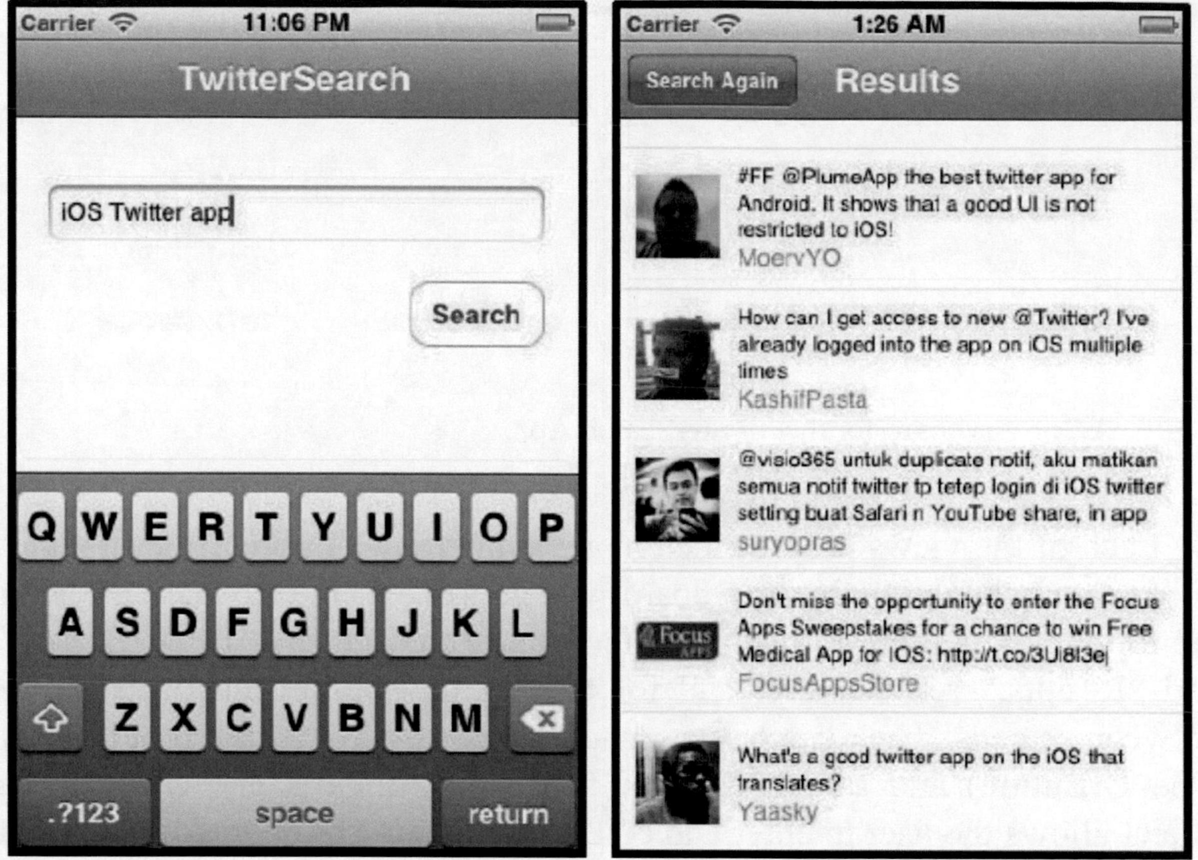

Figure A.1: Search View (Left) and Search Results View (Right).

Figure A.2 shows the storyboard for the TwitterSearchApp. iOS5 developers can create the views which make up their application and specify the transitions between them using a WYSIWYG (What-You-See-Is-What-You-Get) approach. The storyboard graphically depicts the flow of the user interface going from left to right. To specify a transition from one view to another, simply press the Control key and drag your mouse pointer from the first view to the second view. The transition will be shown on the storyboard as a gray arrow joining the two views. This type of transition is known as Segue (pronounced 'Seg-way'). Each of the Segues can be named so that it can be accessed in the code using predefined transition animation effects.

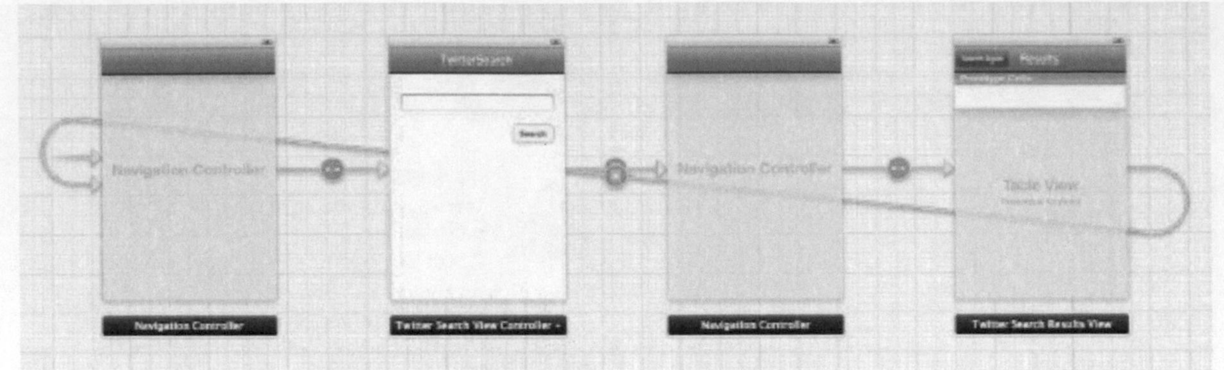

Figure A.2: Storyboard for the TwitterSearch App.

Each of the views is embedded in a Navigation Controller (the first and third scenes shown on the storyboard) which provides the navigation bar on the top which is used to displaying the title of the screen ("TwitterSearch" and "Results," respectively) and for placing navigation buttons to move between screens. The Search View interface consists of a search button (of type UIButton) and a text field (of type UITextField). The UITextField object allows the user to enter and edit text. When a UITextField object has focus, it automatically pops up the virtual keyboard of the device. When the user presses the search button, the application retrieves the contents of the text field and composes an HTTP request containing the search text which is then transmitted to the Twitter server. Once the results arrive, it transitions into the Search Results View which displays the search response in a UITableView object (Figure A.1). This transition is defined graphically using the storyboard as seen in Figure A.2. To define the transition, one needs to press the Control key and drag the mouse pointer from the search button to the next view to the right of the Search View on the storyboard. Similarly for the "Search Again" button in the Search Results View, one defines a transition from this button back to the Search View which is depicted by the curved arrow going from right to left.

In order to access the contents of the text field in the Search View, we define a variable searchTextField of type IBOutlet in TwitterSearchViewController.h (page 79) which is linked to the UITextField object. This link is done graphically using the Storyboard editor in Xcode. IBOutlet variables allow the developer to reference in their code, UI elements visually defined using the Storyboard or Interface Builder.

Since the transition from the Search View to the Search Results View is the trigger for performing the search using Twitter's REST API hence, the actual search operation is done in the TwitterSearchViewController class (refer to TwitterSearchViewController.m on page 80).

TwitterSearchViewController.h

```
#import <UIKit/UIKit.h>
#import <Twitter/Twitter.h>
#import "TwitterSearchResultsViewController.h"

@class TwitterSearchViewResultsController;

@interface TwitterSearchViewController :
UIViewController

@property (strong, nonatomic) IBOutlet UITextField
*searchTextField;

@end
```

iOS provides a function called prepareForSegue() which is triggered whenever there is a transition from one view to another. This function is overridden and implements the main logic for performing the Twitter search and retrieving the results in JSON format. The results are then stored in an array of type NSArray which is then passed on to the TwitterSearchResultsViewController object handling the Search Results View. Finally, the updateTableView method of this object is called to update the table contents of the Search Results View resulting in the user interface which looks something like what is shown in Figure A.1.

TwitterSearchViewController.m

```
// Get reference to the view controller which is of
type
// TwitterSearchResultsViewController
```

```
// Remember that the TwitterSearchResultsView
Controller is
// embedded in a Navigation Controller hence we need
to get
// it's topViewController

TwitterSearchResultsViewController
*searchViewController =
(TwitterSearchResultsViewController *)[[segue
destinationViewController] topViewController];
```

Note how the reference to the TwitterSearchResultsViewController object is retrieved within the prepareForSegue() function.

TwitterSearchResultsViewController.m

```
// Set content and appearance of table view cells.
- (UITableViewCell *)tableView:(UITableView *)
tableView
cellForRowAtIndexPath:(NSIndexPath *)indexPath
{
  static NSString *cellIdentifier = @"Cell";

  UITableViewCell *cell = [tableView
dequeueReusableCellWithIdentifier:cellIdentifier];

  if (cell == nil)
  {
    cell = [[UITableViewCell alloc]
 initWithStyle:UITableViewCellStyleSubtitle
 reuseIdentifier:cellIdentifier];
  }

  // Insert contents of a single Tweet into a cell
  ...
  NSDictionary *tweetMessage = [tweetMessages
 objectAtIndex:[indexPath row]];

  // Set the Tweet contents
  cell.textLabel.text = [tweetMessage objectForKey:
  @"text"];
  cell.textLabel.numberOfLines = 4;
  cell.textLabel.adjustsFontSizeToFitWidth = YES;
```

```
  cell.textLabel.font = [UIFont systemFontOfSize:12];
  cell.textLabel.minimumFontSize = 12;
  cell.textLabel.lineBreakMode =
  UILineBreakModeWordWrap;

  // Set the user name
  cell.detailTextLabel.text = [tweetMessage
objectForKey:@"from_user"];
```

Once the application transitions into the Search Results View, the UITableView object displays the results in a cell-based format where each cell contains a single tweet message along with the name of its author and the author's profile picture. This is done through the cellForRowAtIndexPath() function in TwitterSearchResultsViewController.m which implements the view controller for the Search Results View. The developer must ensure that if the cell being rendered into does not exist (that is, it is equal to *nil*) then it is allocated properly otherwise it will result in a run-time exception.

A.2 ANDROID CODE

An Android application is described in a manifest file (ApplicationMainifest.xml) in the application root directory. This file presents information about the application and describes the components (Activities, Services, BroadcastReceivers and ContentProvider) used in the application. It also declares the permissions required to run the application. For example, our sample application requires network access. android.permission.INTERNET is specified in this file.

ApplicationMainifest.xml

```
<?xml version="1.0" encoding="utf-8"?>
<manifest xmlns:android="http://schemas.android.com/apk/
res/android"
   package="com.morganclaypool.mobile"
   android:versionCode="1"
   android:versionName="1.0">
 <application android:icon="@drawable/icon"
   android:label="@string/app_name">
```

```
  <activity android:name=".TwitterSearchActivity"
        android:label="@string/app_name">
   <intent-filter>
    <action android:name="android.intent.action.MAIN"/>
    <category android:name="android.intent.category.
    LAUNCHER" />
   </intent-filter>
  </activity>
  <activity android:name=".TweetListActivity"></activity>
 </application>
 <uses-permission android:name="android.permission.
 INTERNET"></uses-
permission>
 <uses-sdk android:minSdkVersion="8" />
</manifest>
```

The followings are layout XML code for Search Screen (main.xml), Results Screen (tweetlist.xml), and Tweet View (tweet.xml) on Results Screen.

main.xml

```
<?xml version="1.0" encoding="utf-8"?>
<LinearLayout xmlns:android="http://schemas.android
.com/apk/res/android"
  android:orientation="vertical"
  android:layout_width="fill_parent"
  android:layout_height="fill_parent">
<EditText
  android:id="@+id/keyWordEditText"
  android:layout_width="match_parent"
  android:layout_height="wrap_content"/>
<Button
  android:id="@+id/searchButton"
  android:text="Search"
  android:layout_gravity="center"
  android:layout_width="wrap_content"
  android:layout_height="wrap_content"/>
</LinearLayout>
```

tweetlist.xml

```
<?xml version="1.0" encoding="utf-8"?>
<ListView
xmlns:android="http://schemas.android.com/apk/res
/android"
 android:orientation="vertical"
 android:layout_width="match_parent"
 android:layout_height="match_parent">
</ListView>
```

tweet.xml

```
<?xml version="1.0" encoding="utf-8"?>
<LinearLayout xmlns:android="http://schemas.android
.com/apk/res/android"
  android:layout_width="wrap_content"
  android:layout_height="wrap_content" >
<ImageView
  android:id="@+id/icon"
  android:layout_width="48px"
  android:layout_height="48px"
  android:layout_marginLeft="4px"
  android:layout_marginRight="10px"
  android:layout_marginTop="4px"
  android:layout_marginBottom="4px">
</ImageView>
<TextView
  android:id="@+id/text"
  android:layout_width="wrap_content"
  android:layout_height="wrap_content">
</TextView>
</LinearLayout>
```

One may manually write those human readable XML layout files. Android SDK provides a rich editor for developers to visually design the layout or verify manually edited layout.xml files.

For the search request, we use Apache HttpClient API provided by Android. JSON response message is parsed with JSON Java library from json.org. The Search Screen and Results Screen are defined in TwitterSearch Activity (TwitterSearchActivity.java on page 86) and TweetList Activity (TweetListActivity.java on pages 87–88), respectively.

The screenshots of the demo application are shown in Figure A.3 (page 85), followed by the source code.

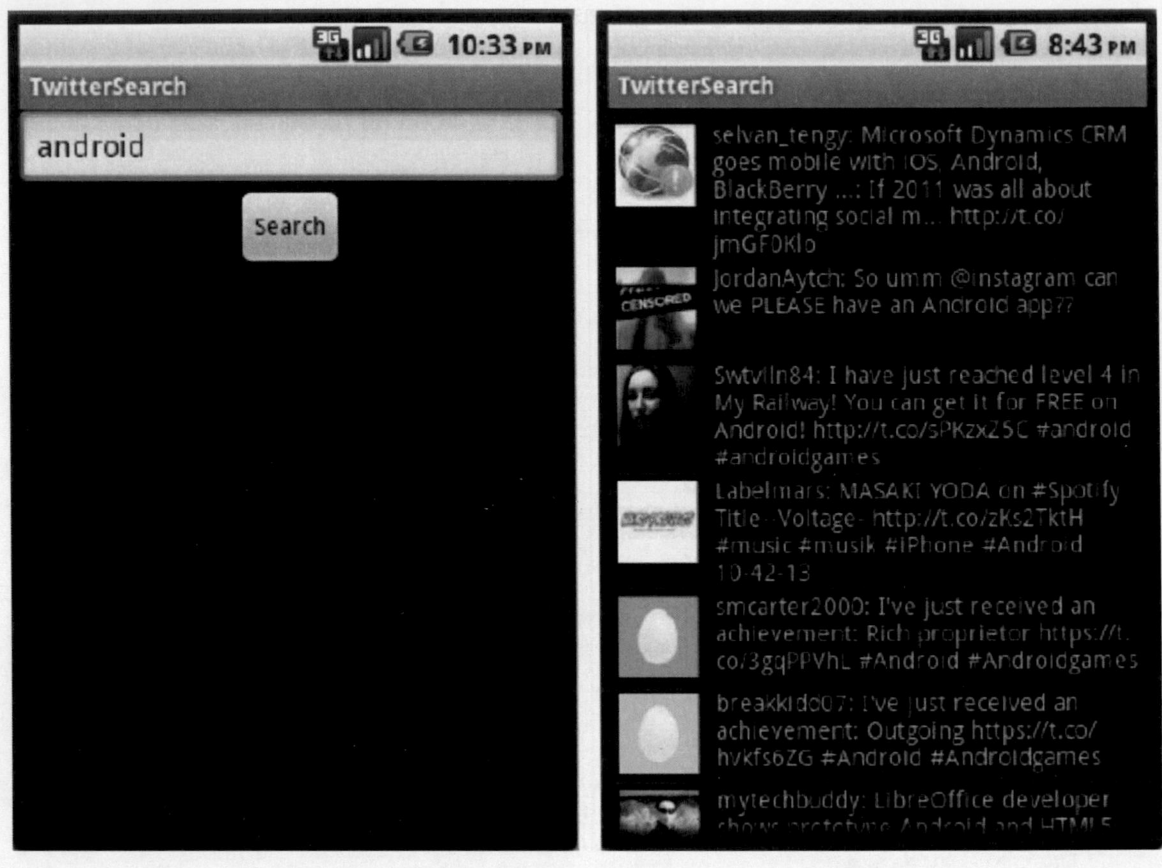

Figure A.3: Android Search/Results Screens for TwitterSearch.

TwitterSearchActivity.java

```
package com.morganclaypool.mobile;

import android.app.Activity;
import android.content.Intent;
import android.os.Bundle;
import android.view.View;
import android.widget.*;

public class TwitterSearchActivity extends Activity {

 @Override
 public void onCreate(Bundle savedInstanceState) {
```

```
  super.onCreate(savedInstanceState);
  setContentView(R.layout.main);

  final Button searchButton = (Button) findViewById
  (R.id.searchButton);
  searchButton.setOnClickListener(new View
  .OnClickListener() {
   public void onClick(View view) {
     Intent intent = new Intent();
     intent.setClass(TwitterSearchActivity.this,
     TweetListActivity.class);
     Bundle bundle = new Bundle();
     final EditText keywordEditText = (EditText)
findViewById(R.id.keyWordEditText);
     bundle.putString("keyword", keywordEditText
     .getText().toString());
     intent.putExtras(bundle);

     startActivity(intent);
   }
  });
 }
```

TweetListActivity.java

```
package com.morganclaypool.mobile;

import java.io.*;
import java.net.*;
import java.util.ArrayList;
import java.util.List;

import org.apache.http.*;
import org.apache.http.client.methods.HttpGet;
import org.apache.http.impl.client.DefaultHttpClient;
import org.apache.http.util.EntityUtils;
import org.json.*;

import android.app.ListActivity;
import android.os.Bundle;

public class TweetListActivity extends ListActivity {

 private static final String TWITTER_SEARCH_API =
```

```
"http://search.twitter.com/search.json?lang=en&q=";
 private DefaultHttpClient httpClient = new
 DefaultHttpClient();

 @Override
 public void onCreate(Bundle savedInstanceState) {
  super.onCreate(savedInstanceState);
  Bundle bundle = this.getIntent().getExtras();
  String keyword = bundle.getString("keyword");
  TweetArrayAdaptor adapter = new TweetArrayAdaptor
  (this,
searchTweets(keyword));
  setListAdapter(adapter);
 }
private List<Tweet> searchTweets(String keyword) {
  List<Tweet> tweetList = new ArrayList<Tweet>();
  String resultString = get(TWITTER_SEARCH_API +
URLEncoder.encode(keyword));
  try{
    JSONArray resultJsonArray = (new
JSONObject(resultString)).getJSONArray("results");
   JSONObject jsonObject = null;
   for (int i = 0; i < resultJsonArray.length(); i++) {
    jsonObject = resultJsonArray.getJSONObject(i);
    Tweet tweet = new Tweet();
    tweet.user = jsonObject.getString("from_user");
    tweet.text = jsonObject.getString("text");
    tweet.iconUrl = jsonObject.getString("profile_image_
    url");
    tweetList.add(tweet);
   }
  } catch (JSONException e) {
   e.printStackTrace();
  }
  return tweetList;
 }

private String get(String url) {
 String responseMessage = null;
 HttpGet httpGet = new HttpGet(url);
 try{
   HttpResponse getResponse = httpClient.execute(httpGet);
   HttpEntity getResponseEntity = getResponse.getEntity();
   if (getResponseEntity != null) {
    responseMessage = EntityUtils.toString(getResponse
    Entity);
   }
  } catch (IOException e) {
```

```
    httpGet.abort();
    e.printStackTrace();
  }
  return responseMessage;
 }
}
```

TweetArrayAdaptor.java

```
package com.morganclaypool.mobile;

import java.io.*;
import java.net.URL;
import java.util.List;

import android.content.Context;
import android.graphics.drawable.Drawable;
import android.view.*;
import android.widget.*;

public class TweetArrayAdaptor extends ArrayAdapter
<Tweet> {

 private final Context context;
 private final List<Tweet> tweets;

 public TweetArrayAdaptor(Context context, List<Tweet>
 tweets) {
  super(context, R.layout.tweet, tweets);
  this.context = context;
  this.tweets = tweets;
 }

 @Override
 public View getView(int position, View convertView,
 ViewGroup parent) {
  LayoutInflater inflater = (LayoutInflater)
context.getSystemService(Context.LAYOUT_INFLATER
_SERVICE);
  Tweet tweet = tweets.get(position);
  View tweetView = inflater.inflate(R.layout.tweet,
  parent, false);
  TextView textView = (TextView) tweetView
  .findViewById(R.id.text);
```

```
  textView.setText(tweet.user + ":" +tweet.text);
  ImageView imageView = (ImageView) tweetView
  .findViewById(R.id.icon);
  imageView.setlmageDrawable(loadlmageFromURL
  (tweet.iconUrl));
  return tweetView;
 }
private Drawable loadlmageFromURL(String url) {
  Drawable drawable = null;
  try {
    InputStream is = (InputStream) new URL(url)
    .getContent();
    drawable = Drawable.createFromStream(is,
    "srcname");
  } catch (lOException e) {
    e.printStackTrace();
  }
  return drawable;
 }
```

Tweet.java

```
package com.morganclaypool.mobile;

public class Tweet {

  public String user;
  public String text;
  public String iconUrl;

}
```

WMAppManifest.xml

```
<?xml version="1.0" encoding="utf-8"?>
<Deployment
xmlns="http://schemas.microsoft.com/windowsphone/2009
/deployment"
AppPlatformVersion="7.0">
 <App xmlns="" ProductID="{68b3e752-40a5-4f0b-a0a7-
 34908b58e9f4}"
Title="TwitterSearch" RuntimeType="Silverlight"
```

```
Version="1.0.0.0"
Genre="apps.normal"  Author="" Description=""
Publisher="">
  <IconPath IsRelative="true"
IsResource="false">ApplicationIcon.png</IconPath>
  <Capabilities>
   <Capability Name="ID_CAP_NETWORKING"/>
  </Capabilities>
  <Tasks>
   <DefaultTask  Name ="_default" NavigationPage=
   "MainPage.xaml"/>
  </Tasks>
  <Tokens>
   <PrimaryToken TokenID="TwitterSearchToken"
   TaskName="_default">
    <TemplateType5>
     <BackgroundImageURI IsRelative="true"
IsResource="false">Background.png</BackgroundImageURI>
     <Count>0</Count>
     <Title>TwitterSearch</Title>
    </TemplateType5>
   </PrimaryToken>
  </Tokens>
 </App>
</Deployment>
```

A.3 WINDOWS PHONE CODE

Windows Phone application metadata is defined in a manifest file (WMAppManifest.xml on page 91). It specifies the capabilities of the application such as networking (ID_CAP_NETWORKING). This manifest file is required to generate the application package (.XAP), the binary build which can be installed to a test enabled Windows Phone device.

App.xaml and the code-behind App.xaml.cs files define the application's entry point, lifecycle, management, application-scoped resources and unhandled exception detection. For our TwitterSearch application, those two files are mostly auto-generated by Windows Phone Silverlight application template with Visual Studio. They are not shown here but can be downloaded with all other files of the Windows Phone code.

Search Screen and Results Screen are defined in MainPage.xaml (pages 92–93) and TweetList.xaml (pages 94–95), respectively. JSON data is parsed and deserialized into objects using the DataContractJsonSerializer class. We are using a technology called Data Binding to link Twitter search result data to UI elements defined in TweetList.xaml. Because of this data binding, the functional programming aspect of C# and C# class properties, the actual code is quite short and condensed.

MainPage.xaml

```
<phone:PhoneApplicationPage
  x:Class="TwitterSearchWindowsPhone.MainPage"
  xmlns="http://schemas.microsoft.com/winfx/2006/xaml
  /presentation"
  xmlns:x="http://schemas.microsoft.com/winfx/2006
  /xaml"
  xmlns:phone="clr-
namespace:Microsoft.Phone.Controls;assembly=
Microsoft.Phone"
  xmlns:shell="clr-namespace:Microsoft.Phone.Shell;
  assembly=Microsoft.Phone"
  xmlns:d="http://schemas.microsoft.com/expression/
  blend/2008"
  xmlns:mc="http://schemas.openxmlformats.org/markup-
  compatibility/2006"
  mc:Ignorable="d" d:DesignWidth="480" d:DesignHeight
  ="768"
  FontFamily="{StaticResource PhoneFontFamilyNormal}"
  FontSize="{StaticResource PhoneFontSizeNormal}"
  Foreground="{StaticResource PhoneForegroundBrush}"
  SupportedOrientations="Portrait" Orientation=
  "Portrait"
  shell:SystemTray.IsVisible="True">
  <Grid x:Name="LayoutRoot" Background="Transparent">
    <Grid.RowDefinitions>
      <RowDefinition Height="Auto"/>
      <RowDefinition Height="*"/>
    </Grid.RowDefinitions>
    <StackPanel x:Name="TitlePanel" Grid.Row="0"
    Margin="12,17,0,28">
      <TextBlock x:Name="ApplicationTitle"
      Text="TWITTER SEARCH"
 Style="{StaticResource PhoneTextNormalStyle}"/>
      <TextBlock x:Name="PageTitle" Text="search"
      Margin="9,-7,0,0"
 Style="{StaticResource PhoneTextTitle1Style}"/>
```

```
    </StackPanel>

    <Grid x:Name="ContentPanel" Grid.Row="1"
    Margin="12,0,12,0">
      <TextBox Height="72" HorizontalAlignment="Left"
      Margin="8,15,0,0"
Name="keywordTextField" Text="" VerticalAlignment=
"Top" Width="460" />
      <Button Content = "Search" Height="72"
      HorizontalAlignment="Left"
Margin="257,84,0,0" Name="searchButton"
VerticalAlignment="Top" Width="160"
Click="searchButton_Click" />
    </Grid>
  </Grid>

</phone:PhoneApplicationPage>
```

TweetList.xaml

```
<phone:PhoneApplicationPage
  x:Class="TwitterSearchWindowsPhone.TweetList"
  xmlns="http://schemas.microsoft.com/winfx/2006/xaml
  /presentation"
  xmlns:x="http://schemas.microsoft.com/winfx/2006
  /xaml"
  xmlns:phone="clr-
namespace:Microsoft.Phone.Controls;assembly=
Microsoft.Phone"
  xmlns:shell="clr-namespace:Microsoft.Phone.Shell;
  assembly=Microsoft.Phone"
  xmlns:d="http://schemas.microsoft.com/expression/
  blend/2008"
  xmlns:mc="http://schemas.openxmlformats.org/
  markup-compatibility/2006"
  FontFamily="{StaticResource PhoneFontFamilyNormal}"
  FontSize="{StaticResource PhoneFontSizeNormal}"
  Foreground="{StaticResource PhoneForegroundBrush}"
  SupportedOrientations="Portrait" Orientation="Portrait"
  mc:Ignorable="d" d:DesignHeight="768"
  d:DesignWidth="480"
  shell:SystemTray.IsVisible="True">
  <Grid x:Name="LayoutRoot" Background="Transparent">
    <Grid.RowDefinitions>
```

```
        <RowDefinition Height="Auto"/>
        <RowDefinition Height="*"/>
    </Grid.RowDefinitions>
    <StackPanel x:Name="TitlePanel" Grid.Row="0" Margin="
    12,17,0,28">
        <TextBlock x:Name="ApplicationTitle" Text="TWITTER
        SEARCH"
Style="{StaticResource PhoneTextNormalStyle}"/>
            <TextBlock  x:Name="PageTitle"  Text="tweets"
Margin="9,
        -7,0,0"
Style="{StaticResource PhoneTextTitle1Style}"/>
    </StackPanel>
    <Grid x:Name="ContentPanel" Grid.Row="1" Margin="12,0,
    12,0">
        <ListBox HorizontalAlignment="Left" Margin="8,8,0,8"
Name="TweetsListBox" Width="460">
            <ListBox.ItemTemplate>
                <DataTemplate>
                    <StackPanel Orientation="Horizontal">
                        <Image Source="{Binding profile_image_url}"
                        Height="73"
Width="73" VerticalAlignment="Top" Margin="0,25,8,0"/>
                        <StackPanel Width="370">
                            <TextBlock Text="{Binding from_user}"
Foreground="#FFC8AB14" FontSize="24"/>
                            <TextBlock Text="{Binding text}" Text
                            Wrapping="Wrap"
FontSize="20"/>
                        </StackPanel>
                    </StackPanel>
                </DataTemplate>
            </ListBox.ItemTemplate>
        </ListBox>
    </Grid>
  </Grid>
</phone:PhoneApplicationPage>
```

MainPage.xaml.cs

```
using System;
using System.Collections.Generic;
using System.Linq;
using System.Net;
```

```
using System.Windows;
using System.Windows.Controls;
using System.Windows.Documents;
using System.Windows.Input;
using System.Windows.Media;
using System.Windows.Media.Animation;
using System.Windows.Shapes;
using Microsoft.Phone.Controls;
using System.Windows.Navigation;

namespace TwitterSearchWindowsPhone
{
  public partial class MainPage :
  PhoneApplicationPage
  {
    public MainPage()
    {
      InitializeComponent();
    }

    private void searchButton_Click(object sender,
    RoutedEventArgs e)
    {
      NavigationService.Navigate(new
Uri(string.Format("/TweetList.xaml?keyword={0}",
keywordTextField.Text),
UriKind.Relative));
    }
  }
}
```

TweetList.xaml.cs

```
using System;
using System.Collections.Generic;
using System.Linq;
using System.Net;
using System.Windows;
using System.Windows.Controls;
using System.Windows.Documents;
using System.Windows.Input;
using System.Windows.Media;
using System.Windows.Media.Animation;
using System.Windows.Shapes;
```

```
using Microsoft.Phone.Controls;
using System.IO;
using System.Runtime.Serialization;
using System.Runtime.Serialization.Json;

namespace TwitterSearchWindowsPhone
{
  public partial class TweetList :
  PhoneApplicationPage
  {

    private WebClient webClient = new WebClient();
    private string keyword = null;

    public TweetList()
    {
      lnitializeComponent();
    }
    protected override void
OnNavigatedTo(System.Windows.Navigation
.NavigationEventArgs e)
    {
      base.OnNavigatedTo(e);
      keyword = this.NavigationContext.QueryString
      ["keyword"];
       webClient.OpenReadCompleted += new
  OpenReadCompletedEventHandler(webClient_OpenRead
  Completed);
       webClient.OpenReadAsync(new
  Uri("http://search.twitter.com/search.json?&lang
  =en&q=" + keyword));
       }

       void webClient_OpenReadCompleted(object sender,
  OpenReadCompletedEventArgs e)
       {
         try
         {
           if (e.Error != null)
           {
             return;
           }
           DataContractJsonSerializer
           dataContractJsonSerializer = new
  DataContractJsonSerializer(typeof(Tweets));
           Tweets tweets =
  (Tweets)dataContractJsonSerializer.ReadObject
  (e.Result);
```

```
        TweetsListBox.ItemsSource = from tweet in
        tweets.results
                                 select new Tweet
                                 {
                                   from_user =
                                   tweet.from_user,
                                   text = tweet.text,
                                   profile_image_url =
                                   tweet.profile_image_url
                                 };
      }
      catch {}
    }
  }
  public class Tweet
  {
    public string text {get; set;}
    public string from_user {get; set;}
    public string profile_image_url { get; set; }
  }

  public class Tweets
  {
    public Tweet[] results { get; set; }
  }

}
```

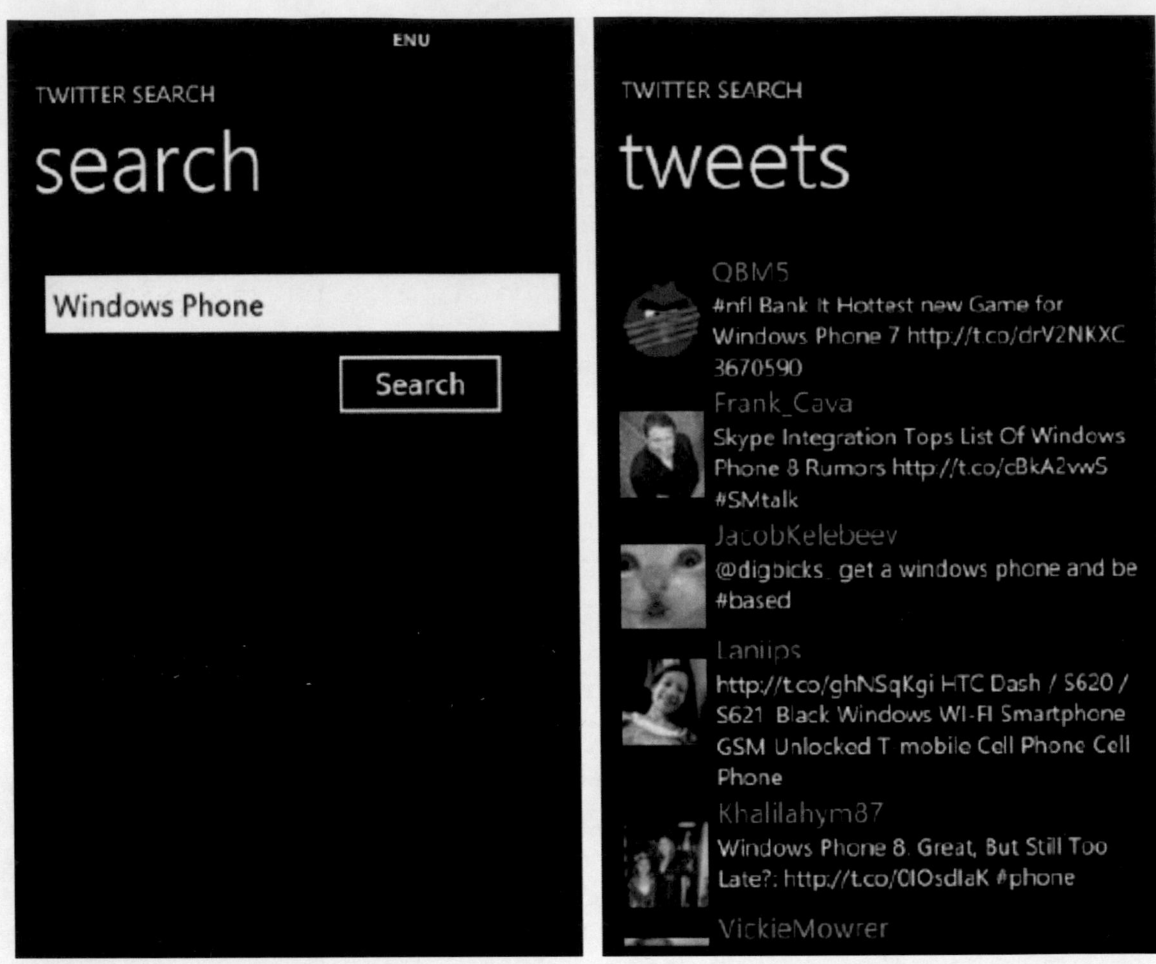

Figure A.4: Windows Phone Search/Results Screens for TwitterSearch.

A.4 MOBILE WEB CODE

The Mobile Web version of this TwitterSearch application has three files: index.html, twitter.css (on page 101) and twitter.js (on pages 102–103). The HTML file includes two <div> in the HTML body: searchForm and tweetList. The CSS file is nothing fancy but gives the basic styles to different elements on the HTML file. JavaScript is used to dynamically manipulate the DOM (adding search results into tweetList div), and to show/hide one of the two div's to give the perception of two separate screens.

index.html

```
<!DOCTYPE html>
<html>
 <head>
  <title>Twitter Search</title>
  <link rel="stylesheet" href="twitter.css" type=
  "text/css" />
  <script src="twitter.js" type="text/javascript">
  </script>
  <meta name="viewport" content="user-scalable=no,
  width=device-width,
initial-scale=1.0, maximum-scale=1.0"/>
  <link rel="apple-touch-icon" href="twitter.png" />
  <meta name="apple-mobile-web-app-capable" content=
  "yes" />
  <meta name="apple-mobile-web-app-status-bar-style"
  content="black" />
 </head>
 <body>
  <h1>TwitterSearch</h1>
  <div id="searchForm">
   <form onsubmit="searchTweets(); return false;">
    <input type="text" id="keywordTextbox"/>
    <br/>
    <input type="submit" value="Search" id=
    "searchButton"/>
   </form>
  </div>
  <div id="tweetList">
  </div>
 </body>
```

twitter.css

```
body {
 font-family: Arial,sans-serif;
}
h1 {
 font-size: 1.0em;
 padding: 4px;
}
td {
  padding: 4px;
}
.tweet {
```

```
 vertical-align: top;
}
.icon {
  vertical-align: text-top;
}
input[type = text] {
  border: 1px solid black;
  width: 95%;
  margin: 4px;
  padding:4px;
}
input[type = submit] {
  border: 1px solid white;
  -webkit-box-shadow:0 0 4px #333333;
  box-shadow:0 0 4px #333333;
  width:99%;
  padding:2px;
  margin: 4px;
}
```

twitter.js

```
function searchTweets(){
  var query =
"http://search.twitter.com/search.json?callback=
showResults&q=";
  query += document.getElementById("keywordTextbox")
  .value;
  var script = document.createElement("script");
  script.src = query;
  document.getElementsByTagName("head")[0].appendChild
  (script);
}

function showResults(response){
  var tweets = response.results;
  var rows = tweets.map(function(tweet){
    return createTweetRow(tweet.from_user, tweet
    .profile_image_url,
tweet.text);
  });
  document.getElementById("tweetList").innerHTML =
  "<input type='submit'
value='Back' onclick='showSearchForm();' /><br />
```

```
<table
id='resultsTable'></table>";
  rows.forEach(function(row){
    document.getElementById("resultsTable")
    .appendChild(row);
  });
  document.getElementById("searchForm").style.display
  = "none";
  document.getElementById("tweetList").style.display
  = "block";
}

function showSearchForm() {
  document.getElementById("searchForm").style.display
  = "block";
  document.getElementById("tweetList").style.display
  = "none";
}

function createTweetRow(user, iconUrl, tweet){
  var tweetRow = document.createElement("tr");
  var iconCell = document.createElement("td");
  iconCell.setAttribute("class", "icon");
  var icon = document.createElement("img");
  icon.src = iconUrl;
  icon.setAttribute("alt", user);
  icon.setAttribute("height", 48);
  icon.setAttribute("width", 48);
  iconCell.appendChild(icon);
  tweetRow.appendChild(iconCell);
  var tweetCell = document.createElement("td");
  tweetCell.setAttribute("class", "tweet");
  tweetCell.appendChild(document.createTextNode
  (user + ": " + tweet));
  tweetRow.appendChild(tweetCell);
  return tweetRow;
}
```

Figure A.5 (next page) shows this mobile Web application running on the iPhone Safari Browser.

Figure A.6 (on page 105) shows the same TwitterSearch application bookmarked in the Home screen and running with full screen mode on the iPhone emulator. This is achieved by adding the following lines in the HTML code:

```
<link rel="apple-touch-icon" href="twitter.png" />
<meta name="apple-mobile-web-app-capable" content="yes" />
<meta name="apple-mobile-web-app-status-bar-style"
content="black" />
```

Those additional link and meta attributes are introduced by iOS but they are also supported by Android. The first apple-touch-icon link will add that icon to the application grid when the user bookmarks this web application to the home screen. The "apple-mobile-web-app-capable" <meta> tag makes a web application run in full-screen mode by removing the default address bar and navigation buttons of the browser. The "apple-mobile-web-app-status-bar-style" <meta> tag shows the status bar with a black background.

If this application is to be made offline, an AppCache manifest file should be added to the list of all files that the browser should cache.

```
CACHE MANIFEST
# Version 1.0
CACHE:
index.html
```

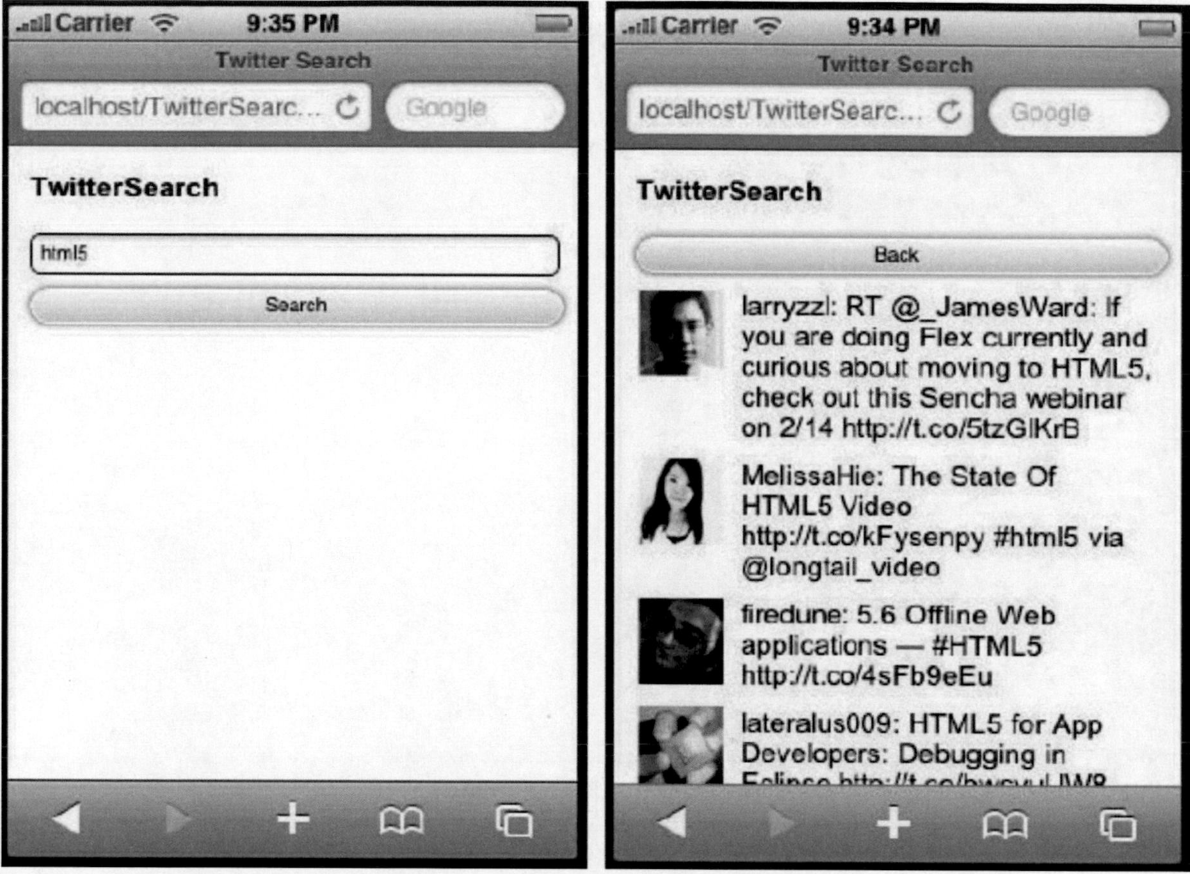

Figure A.5: Mobile Web Search/Results Screens for TwitterSearch.

```
twitter.css
twitter.js
twitter.png
NETWORK:
http://search.twitter.com
```

Then, a link to that manifest file should be added in the HTML file as shown below.

```
<html manifest="manifest.mf">
```

Figure A.6: Full Screen Mobile Web App with Launching Icon on iPhone.

Table A.1: Some Useful Web References.

TwitterSearch API	`https://dev.twitter.com/docs/api/1/get/search`
JSON	`http://www.json.org/`
JSON in Java	`http://json.org/java/`
Android: Binding to Data with AdapterView	`http://developer.android.com/guide/topics/ui/binding.html`
Silverlight: Displaying	`http://msdn.microsoft.com/en-us/library/gg680271(v=pandp.11).aspx`

Data with Data Binding	
C# Properties	http://msdn.microsoft.com/en-us/library/x9fsa0sw.aspx
JSONP	http://en.wikipedia.org/wiki/JSONP
A Beginner's Guide to Using the Application Cache	http://www.html5rocks.com/en/tutorials/appcache/beginner/

Table A.2: TwitterSearch Sample App Download Links.

iOS	http://www.icta.ufl.edu/mobileplatforms/iOS_TwitterSearch_App.zip
Android	http://www.icta.ufl.edu/mobileplatforms/TwitterSearchAndroid.zip
Windows Phone	http://www.icta.ufl.edu/mobileplatforms/TwitterSearchWindowsPhone.zip
Mobile Web	http://www.icta.ufl.edu/mobileplatforms/TwitterSearchMobileWeb.zip

Authors' Biographies

SUMI HELAL

Sumi Helal, Ph.D. is a Professor of Computer and Information Science and Engineering (CISE) at the University of Florida. His research interests span the areas of Pervasive Computing, Mobile Computing and networking and Internet Computing. He directs the Mobile and Pervasive Computing Laboratory at the CISE department, and is co-founder and director of the Gator Tech Smart House—an experimental home for applied research in the domain of aging, disability and independence. Dr. Helal is the Associate Editor in Chief of IEEE Computer and an editorial board member of IEEE Pervasive Computing. Contact him at: `sumi.helal@gmail.com`

RAJA BOSE

Raja Bose, Ph.D. is a senior researcher at Nokia Research Center North America Lab in Palo Alto, CA, USA. His research interests focus on the theme of mobile interoperability, involving the creation of new systems and services that provide innovative user experiences by leveraging the interoperability of mobile devices with other smart devices and environments. He received a PhD in computer engineering from University of Florida. Contact him at: raja.bose@nokia.com

WENDONG LI

Wendong Li, ME is a senior software engineer at Nokia Burlington, MA, USA. He has 10 years of experience in mobile software development, starting from the early days of Java ME to the latest smart phones. His current work involves mobile, social and cloud. He received a Master's degree in Computer Engineering from the University of Florida. Contact him at: wendong.li@gmail.com

Printed in the United States
by Baker & Taylor Publisher Services